Cram101 Textbook Outlines to accompany:

Feedback Control of Dynamic Systems

Gene F. Franklin, 6th Edition

Learning System

Cram101 Textbook Outlines is a learning system. The notes in this book are the highlights of your textbook, you will never have to highlight a book again.

How to use this book. Take this book to class, it is your notebook for the lecture. The notes and highlights on the left hand side of the pages follow the outline and order of the textbook. All you have to do is follow along while your instructor presents the lecture. Circle the items emphasized in class and add other important information on the right side. With Cram101 Textbook Outlines you'll spend less time writing and more time listening. Learning becomes more efficient.

Cram101.com Online

Increase your studying efficiency by using Cram101.com's practice tests and online reference material. It is the perfect complement to Cram101 Textbook Outlines. Use self-teaching matching tests or simulate in-class testing with comprehensive multiple choice tests, or simply use Cram's true and false tests for quick review. Cram101.com even allows you to enter your in-class notes for an integrated studying format combining the textbook notes with your class notes.

Visit **www.Cram101.com**, click Sign Up at the top of the screen, and enter **DK73DW12617** in the promo code box on the registration screen. Your access to www.Cram101.com is discounted by 50% because you have purchased this book. Sign up and stop highlighting textbooks forever.

 ISBN(s): 9781614902232. PUBE-4 .201153

Feedback Control of Dynamic Systems
Gene F. Franklin, 6th

CONTENTS

Reference	Hydrogen silsesquioxane is a negative tone electron-beam resist. Thin HSQ layers have been used to show very high resolution (~10 nm) feature sizes. Reference · http://cat.inist.fr/?aModele=afficheN'cpsidt=18807320 :'.
Stability	In probability theory, the stability of a random variable is the property that a linear combination of two independent copies of the variable has the same distribution, up to location and scale parameters. The distributions of random variables having this property are said to be "stable distributions". Results available in probability theory show that all possible distributions having this property are members of a four-parameter family of distributions.
Step response	The step response of a system in a given initial state consists of the time evolution of its outputs when its control inputs are Heaviside step functions. In electronic engineering and control theory, step response is the time behaviour of the outputs of a general system when its inputs change from zero to one in a very short time. The concept can be extended to the abstract mathematical notion of a dynamical system using an evolution parameter.
Comparator	In electronics, a comparator is a device that compares two voltages or currents and switches its output to indicate which is larger. Input voltage range The input voltages must stay within the limits specified by the manufacturer. Early integrated comparators, like the LM111 family, and certain high-speed comparators like the LM119 family, require input voltage ranges substantially lower than the power supply voltages (±15 V vs. 36V).

Chapter 1. An Overview and Brief History of Feedback Control

Controller	In control theory, a controller is a device which monitors and affects the operational conditions of a given dynamical system. The operational conditions are typically referred to as output variables of the system which can be affected by adjusting certain input variables. For example, the heating system of a house can be equipped with a thermostat (controller) for sensing air temperature (output variable) which can turn on or off a furnace or heater when the air temperature becomes too low or too high.
Block	A block in the context of extraction of petroleum, (and natural gas) signifies a sub-division of an underground reservoir for the purpose of administering a certain production allowable.
Block diagram	Block diagram is a diagram of a system, in which the principal parts or functions are represented by blocks connected by lines, that show the relationships of the blocks. They are heavily used in the engineering world in hardware design, software design, and process flow diagrams. The block diagram is typically used for a higher level, less detailed description aimed more at understanding the overall concepts and less at understanding the details of implementation.
Float	Liquid level floats, also known as float balls, are spherical, cylindrical, oblong or similarly-shaped objects, made from either rigid or flexible material, that are buoyant in water and other liquids. They are non-electrical hardware frequently used as visual sight-indicators for surface demarcation and level measurement. They may also be incorporated into switch mechanisms or translucent fluid-tubes as a component in monitoring or controlling liquid level.
Liquid	Liquid is one of the three classical states of matter. Like a gas, a liquid is able to flow and take the shape of a container. Some liquids resist compression, while others can be compressed.
Valve	A valve is a device that regulates the flow of a fluid (gases, liquids, fluidized solids, or slurries) by opening, closing, or partially obstructing various passageways. Valves are technically pipe fittings, but are usually discussed as a separate category. In an open valve, fluid flows in a direction from higher pressure to lower pressure.
Steam	Steam is either mist , or the gas phase of water (water vapor).

In common speech, steam most often refers to the visible white mist that condenses above boiling water as the hot vapor mixes with the cooler air. This mist consists of tiny droplets of liquid water.

Steam engine

A steam engine is a heat engine that performs mechanical work using steam as its working fluid.

The idea of using boiling water to produce mechanical motion has a long history, going back about 2,000 years. Early devices were not practical power producers, but more advanced designs producing usable power have become a major source of mechanical power over the last 300 years, beginning with applications for removing water from mines using vacuum engines.

Governor

A governor, is a device used to measure and regulate the speed of a machine, such as an engine. A classic example is the centrifugal governor, also known as the Watt or fly-ball governor, which uses weights mounted on spring-loaded arms to determine how fast a shaft is spinning, and then uses proportional control to regulate the shaft speed.

History

Centrifugal governors were used to regulate the distance and pressure between millstones in windmills since the 17th century.

Frequency

Frequency is the number of occurrences of a repeating event per unit time. It is also referred to as temporal frequency. The period is the duration of one cycle in a repeating event, so the period is the reciprocal of the frequency.

Frequency response

Frequency response is the measure of any system's output spectrum in response to an input signal. In the audible range it is usually referred to in connection with electronic amplifiers, microphones and loudspeakers. Radio spectrum frequency response can refer to measurements of coaxial cables, category cables, video switchers and wireless communications devices.

Regulator	In automatic control, a regulator is a device which has the function of maintaining a designated characteristic. It performs the activity of managing or maintaining a range of values in a machine. The measurable property of a device is managed closely by specified conditions or an advance set value; or it can be a variable according to a predetermined arrangement scheme.

Chapter 2. Dynamic Models

Dynamics

In the field of physics, the study of the causes of motion and changes in motion is dynamics. In other words the study of forces and why objects are in motion. Dynamics includes the study of the effect of torques on motion.

System identification

In control engineering, the field of system identification uses statistical methods to build mathematical models of dynamical systems from measured data. System identification also includes the optimal design of experiments for efficiently generating informative data for fitting such models.

Overview

A dynamical mathematical model in this context is a mathematical description of the dynamic behavior of a system or process in either the time or frequency domain.

Motion

In physics, motion is a change in position of an object with respect to time. Change in action is the result of an unapplied force. Motion is typically described in terms of velocity, acceleration, displacement, and time.

Step response

The step response of a system in a given initial state consists of the time evolution of its outputs when its control inputs are Heaviside step functions. In electronic engineering and control theory, step response is the time behaviour of the outputs of a general system when its inputs change from zero to one in a very short time. The concept can be extended to the abstract mathematical notion of a dynamical system using an evolution parameter.

Suspension

In topology, the suspension SX of a topological space X is the quotient space:

$$SX = (X \times I)/\{(x_1, 0) \sim (x_2, 0) \text{ and } (x_1, 1) \sim (x_2, 1) \text{ for all } x_1$$

of the product of X with the unit interval I = [0, 1]. Intuitively, we make X into a cylinder and collapse both ends to two points. One views X as "suspended" between the end points.

Chapter 2. Dynamic Models

Free body	Free body is the generic term used by physicists and engineers to describe an object--be it a bowling ball, a spacecraft, pendulum, a television, or anything else--which can be considered as moving as a single unit. The object doesn't have to be "free" in the usual sense of the word--it could be completely prevented from going anywhere, or it could be trapped in an orbit. To study the stresses acting in a material, small elements or sections of beams are treated as free bodies.
Schematic	A Schematic is a diagram that represents the elements of a system using abstract, graphic symbols rather than realistic pictures. A Schematic usually omits all details that are not relevant to the information the Schematic is intended to convey, and may add unrealistic elements that aid comprehension. For example, a subway map intended for riders may represent a subway station with a dot; the dot doesn`t resemble the actual station at all but gives the viewer information without unnecessary visual clutter.
Mechanism	A mechanism is a combination of resistant bodies, so interconnected that by applying force or motion to one or more of those bodies, some of those bodies are caused to perform desired work accompanied by desired motions. Mechanisms are used to convert one type of motion into another. The connections present in a mechanism are called kinematic pairs.

Pendulum

Derivations from Pendulum.

Analysis of a simple gravity pendulum

To begin, we shall make three assumptions about the simple pendulum:

- The rod/string/cable on which the bob is swinging is massless, does not stretch, and always remains taut.
- The bob is a point mass.
- Motion occurs in a 2-dimensional plane, i.e. pendulum does not swing into and out of the page.

Consider Figure 1, showing the forces acting on a simple pendulum. Note that the path of the pendulum sweeps out an arc of a circle.

Simulink

Simulink, developed by MathWorks, is a commercial tool for modeling, simulating and analyzing multidomain dynamic systems. Its primary interface is a graphical block diagramming tool and a customizable set of block libraries. It offers tight integration with the rest of the MATLAB environment and can either drive MATLAB or be scripted from it.

Nonlinear

In mathematics, a nonlinear system is a system which is not linear, that is, a system which does not satisfy the superposition principle, a nonlinear system is any problem where the variable(s) to be solved for cannot be written as a linear combination of independent components. A nonhomogeneous system, which is linear apart from the presence of a function of the independent variables, is nonlinear according to a strict definition, but such systems are usually studied alongside linear systems, because they can be transformed to a linear system of multiple variables.

Translation

In physics, translation is movement that changes the position of an object, as opposed to rotation. For example, according to Whittaker:

If a body is moved from one position to another, and if the lines joining the initial and final points of each of the points of the body are a set of parallel straight lines of length l, so that the orientation of the body in space is unaltered, the displacement is called a translation parallel to the direction of the lines, through a distance l.

- E.T. Whittaker: A Treatise on the Analytical Dynamics of Particles and Rigid Bodies, p. 1

A translation is the operation changing the positions of all points (x, y, z) of an object according to the formula

$$(x, y, z) \to (x + \Delta x, y + \Delta y, z + \Delta z)$$

where $(\Delta x, \Delta y, \Delta z)$ is the same vector for each point of the object.

Inverted pendulum	An inverted pendulum is a pendulum which has its mass above its pivot point. It is often implemented with the pivot point mounted on a cart that can move horizontally and may be called a cart and pole. Whereas a normal pendulum is stable when hanging downwards, an inverted pendulum is inherently unstable, and must be actively balanced in order to remain upright, either by applying a torque at the pivot point or by moving the pivot point horizontally as part of a feedback system.
Parameter	In mathematics, statistics, and the mathematical sciences, a Parameter is a quantity that serves to relate functions and variables using a common variable (often t) when such a relationship would be difficult to explicate with an equation. In different contexts the term may have special uses. · In a section on frequently misused words in his book The Writer's Art, James J. Kilpatrick quoted a letter from a correspondent, giving examples to illustrate the correct use of the word Parameter: · A parametric equaliser is an audio filter that allows the frequency of maximum cut or boost to be set by one control, and the size of the cut or boost by another. These settings, the frequency level of the peak or trough, are two of the Parameters of a frequency response curve, and in a two-control equaliser they completely describe the curve.
Development	In classical differential geometry, development refers to the simple idea of rolling one smooth surface over another in Euclidean space. For example, the tangent plane to a surface (such as the sphere or the cylinder) at a point can be rolled around the surface to obtain the tangent-plane at other points. The tangential contact between the surfaces being rolled over one another provides a relation between points on the two surfaces.

Amplifier	Generally, an Amplifier or simply amp, is any device that changes, usually increases, the amplitude of a signal. The relationship of the input to the output of an Amplifier--usually expressed as a function of the input frequency--is called the transfer function of the Amplifier, and the magnitude of the transfer function is termed the gain. In popular use, the term usually describes an electronic Amplifier, in which the input 'signal' is usually a voltage or a current.
Differential	A differential is a device, usually but not necessarily employing gears, capable of transmitting torque and rotation through three shafts, almost always used in one of two ways: in one way, it receives one input and provides two outputs--this is found in most automobiles--and in the other way, it combines two inputs to create an output that is the sum, difference, or average, of the inputs. In automobiles and other wheeled vehicles, the differential allows each of the driving roadwheels to rotate at different speeds, while for most vehicles supplying equal torque to each of them. Purpose A vehicle's wheels rotate at different speeds, mainly when turning corners.
Voltage	The voltage between two points is a short name for the electrical force that would drive an electric current between those points. Specifically, voltage is equal to energy per unit charge. In the case of static electric fields, the voltage between two points is equal to the electrical potential difference between those points.
Capacitor	A capacitor is a passive electronic component consisting of a pair of conductors separated by a dielectric (insulator). When there is a potential difference (voltage) across the conductors, a static electric field develops in the dielectric that stores energy and produces a mechanical force between the conductors. An ideal capacitor is characterized by a single constant value, capacitance, measured in farads.

Current	In mathematics, more particularly in functional analysis, differential topology, and geometric measure theory, a k-current in the sense of Georges de Rham is a functional on the space of compactly supported differential k-forms, on a smooth manifold M. Formally currents behave like Schwartz distributions on a space of differential forms. In a geometric setting, they can represent integration over a submanifold, generalizing the Dirac delta function, or more generally even directional derivatives of delta functions (multipoles) spread out along subsets of M. Let $\Omega_c^m(\mathbb{R}^n)$ denote the space of smooth m-forms with compact support on R^n. A current is a linear functional on $\Omega_c^m(\mathbb{R}^n)$ which is continuous in the sense of distributions.
Inductor	An inductor is a passive electrical component that can store energy in a magnetic field created by the electric current passing through it. An inductor's ability to store magnetic energy is measured by its inductance, in units of henries. Typically an inductor is a conducting wire shaped as a coil; the loops help to create a strong magnetic field inside the coil due to Ampere's Law.
Frequency	Frequency is the number of occurrences of a repeating event per unit time. It is also referred to as temporal frequency. The period is the duration of one cycle in a repeating event, so the period is the reciprocal of the frequency.
Frequency response	Frequency response is the measure of any system's output spectrum in response to an input signal. In the audible range it is usually referred to in connection with electronic amplifiers, microphones and loudspeakers. Radio spectrum frequency response can refer to measurements of coaxial cables, category cables, video switchers and wireless communications devices.
Transfer function	A transfer function is a mathematical representation, in terms of spatial or temporal frequency, of the relation between the input and output of a linear time-invariant system. With optical imaging devices, for example, it is the Fourier transform of the point spread function (hence a function of spatial frequency) i.e. the intensity distribution caused by a point object in the field of view. Explanation

Transfer functions are commonly used in the analysis of systems such as single-input single-output filters, typically within the fields of signal processing, communication theory, and control theory.

Electromagnet

An electromagnet is a type of magnet whose magnetic field is produced by the flow of electric current. The magnetic field disappears when the current ceases. Electromagnets are very widely used as components of other electrical devices, such as motors, generators, relays, loudspeakers, hard disks, MRI machines, scientific instruments, and magnetic separation equipment, as well as being employed as industrial lifting electromagnets for picking up and moving heavy iron objects like scrap iron.

DC motor

A DC motor is an electric motor that runs on direct current (DC) electricity.

Brushed

The brushed DC electric motor generates torque directly from DC power supplied to the motor by using internal commutation, stationary permanent magnets, and rotating electrical magnets. Like all electric motors or generators, torque is produced by the principle of Lorentz force, which states that any current-carrying conductor placed within an external magnetic field experiences a torque or force known as Lorentz force.

Compensation

In engineering, compensation is planning for side effects or other unintended issues in a design. The design of an invention can itself also be to compensate for some other existing issue or exception.

One example is in a voltage-controlled crystal oscillator (VCXO), which is normally affected not only by voltage, but to a lesser extent by temperature.

Armature

In electrical engineering, an armature generally refers to one of the two principal electrical components of an electromechanical machine-generally in a motor or generator, but it may also mean the pole piece of a permanent magnet or electromagnet, or the moving iron part of a solenoid or relay. The other component is the field winding or field magnet. The role of the "field" component is simply to create a magnetic field (magnetic flux) for the armature to interact with, so this component can comprise either permanent magnets, or electromagnets formed by a conducting coil.

Rotor

The rotor is the non-stationary part of a rotary electric motor, electric generator or alternator, which rotates because the wires and magnetic field of the motor are arranged so that a torque is developed about the rotor's axis. In some designs, the rotor can act to serve as the motor's armature, across which the input voltage is supplied. The stationary part of an electric motor is the stator.

Curve

In mathematics, a Curve consists of the points through which a continuously moving point passes. This notion captures the intuitive idea of a geometrical one-dimensional object, which furthermore is connected in the sense of having no discontinuities or gaps. Simple examples include the sine wave as the basic Curve underlying simple harmonic motion, and the parabola.

Fluid

In physics, a fluid is a substance that continually deforms (flows) under an applied shear stress, no matter how small. Fluids are a subset of the phases of matter and include liquids, gases, plasmas and, to some extent, plastic solids.

In common usage, "fluid" is often used as a synonym for "liquid", with no implication that gas could also be present.

Room temperature

Room temperature is a common term that can either denote a certain temperature to which humans are accustomed or a specific temperature. In the United Kingdom for instance, room temperature is traditionally 18 °C (64 °F) (ironically this is nowadays regarded as being too cool, possibly due to the widespread use of central heating). Room temperature may otherwise be indicated by general human comfort, with the common range of 20 °C (68 °F) to 25 °C (78 °F), though people may become acclimatized to higher or lower temperatures.

Thermal conductivity

In physics, thermal conductivity, k, is the property of a material describing its ability to conduct heat. It appears primarily in Fourier's Law for heat conduction. Thermal conductivity is measured in watts per kelvin-metre ($W \cdot K^{-1} \cdot m^{-1}$, i.e. W/(K·m).

Conductivity	The conductivity of an electrolyte solution is a measure of its ability to conduct electricity. The SI unit of conductivity is siemens per meter (S/m). Conductivity measurements are used routinely in many industrial and environmental applications as a fast, inexpensive and reliable way of measuring the ionic content in a solution.
Resistance	"Resistance" as initially used by Sigmund Freud, referred to patients blocking memories from conscious memory. This was a key concept, since the primary treatment method of Freud's talk therapy required making these memories available to the patient's consciousness. "Resistance" expanded Later, Freud described five different forms of resistance.
Temperature	Temperature is a physical property of matter that quantitatively expresses the common notions of hot and cold. Objects of low temperature are cold, while various degrees of higher temperatures are referred to as warm or hot. Quantitatively, temperature is measured with thermometers, which may be calibrated to a variety of temperature scales.
Heat exchanger	A heat exchanger is a device built for efficient heat transfer from one medium to another. The media may be separated by a solid wall, so that they never mix, or they may be in direct contact. They are widely used in space heating, refrigeration, air conditioning, power plants, chemical plants, petrochemical plants, petroleum refineries, natural gas processing, and sewage treatment.
Hydraulics	Hydraulics is a topic in applied science and engineering dealing with the mechanical properties of liquids. Fluid mechanics provides the theoretical foundation for hydraulics, which focuses on the engineering uses of fluid properties. In fluid power, hydraulics is used for the generation, control, and transmission of power by the use of pressurized liquids.

Linearization	In mathematics and its applications, linearization refers to finding the linear approximation to a function at a given point. In the study of dynamical systems, linearization is a method for assessing the local stability of an equilibrium point of a system of nonlinear differential equations or discrete dynamical systems. This method is used in fields such as engineering, physics, economics, and ecology.
Perspective	Perspective, in context of vision and visual perception, is the way in which objects appear to the eye based on their spatial attributes; or their dimensions and the position of the eye relative to the objects. There are two main meanings of the term: linear perspective and aerial perspective. Linear perspective As objects become more distant they appear smaller because their visual angle decreases.

Laplace transform	In mathematics, the Laplace transform is a widely used integral transform. Denoted $\mathcal{L}\{f(t)\}$, it is a linear operator of a function f(t) with a real argument t (t ≥ 0) that transforms it to a function F(s) with a complex argument s. This transformation is essentially bijective for the majority of practical uses; the respective pairs of f(t) and F(s) are matched in tables.
Perspective	Perspective, in context of vision and visual perception, is the way in which objects appear to the eye based on their spatial attributes; or their dimensions and the position of the eye relative to the objects. There are two main meanings of the term: linear perspective and aerial perspective. Linear perspective As objects become more distant they appear smaller because their visual angle decreases.
Dynamics	In the field of physics, the study of the causes of motion and changes in motion is dynamics. In other words the study of forces and why objects are in motion. Dynamics includes the study of the effect of torques on motion.
Convolution	In mathematics and, in particular, functional analysis, Convolution is a mathematical operation on two functions f and g, producing a third function that is typically viewed as a modified version of one of the original functions. Convolution is similar to cross-correlation. It has applications that include statistics, computer vision, image and signal processing, electrical engineering, and differential equations.
Signal	In the fields of communications, signal processing, and in electrical engineering more generally, a signal is any time-varying or spatial-varying quantity. In the physical world, any quantity measurable through time or over space can be taken as a signal. Within a complex society, any set of human information or machine data can also be taken as a signal.

Impulse

In classical mechanics, an impulse is defined as the integral of a force with respect to time. When a force is applied to a rigid body it changes the momentum of that body. A small force applied for a long time can produce the same momentum change as a large force applied briefly, because it is the product of the force and the time for which it is applied that is important.

Pulse

In signal processing, the term pulse has the following meanings:

1. A rapid, transient change in the amplitude of a signal from a baseline value to a higher or lower value, followed by a rapid return to the baseline value.
2. A rapid change in some characteristic of a signal, e.g., phase or frequency, from a baseline value to a higher or lower value, followed by a rapid return to the baseline value.

Pulse shapes

Pulse shapes can arise out of a process called pulse-shaping. Optimum pulse shape depends on the application.

Rectangular pulse

These can be found in pulse waves, square waves, boxcar functions, and rectangular functions.

Impulse response

In signal processing, the impulse response, of a dynamic system is its output when presented with a brief input signal, called an impulse. More generally, an impulse response refers to the reaction of any dynamic system in response to some external change. In both cases, the impulse response describes the reaction of the system as a function of time (or possibly as a function of some other independent variable that parameterizes the dynamic behavior of the system).

Property

In modern philosophy, mathematics, and logic, a property is an attribute of an object; a red object is said to have the property of redness. The property may be considered a form of object in its own right, able to possess other properties. It differs from the logical concept of class by not having any concept of extensionality, and from the philosophical concept of class in that a property is considered to be distinct from the objects which possess it.

Integral

Integration is an important concept in mathematics and, together with differentiation, is one of the two main operations in calculus. Given a function f of a real variable x and an interval [a, b] of the real line, the definite integral

is defined informally to be the net signed area of the region in the xy-plane bounded by the graph of f, the x-axis, and the vertical lines x = a and x = b.

The principles of integration were formulated independently by Isaac Newton and Gottfried Leibniz in the late 17th century.

Frequency

Frequency is the number of occurrences of a repeating event per unit time. It is also referred to as temporal frequency. The period is the duration of one cycle in a repeating event, so the period is the reciprocal of the frequency.

Frequency response

Frequency response is the measure of any system's output spectrum in response to an input signal. In the audible range it is usually referred to in connection with electronic amplifiers, microphones and loudspeakers. Radio spectrum frequency response can refer to measurements of coaxial cables, category cables, video switchers and wireless communications devices.

Ratio

In mathematics, a ratio expresses the magnitude of quantities relative to each other. Specifically, the ratio of two quantities indicates how many times the first quantity is contained in the second and may be expressed algebraically as their quotient. Mathematically, a proportion is defined as the equality of two ratios.

Semiconductor

A semiconductor is a material with electrical conductivity due to electron flow (as opposed to ionic conductivity) intermediate in magnitude between that of a conductor and an insulator. This means a conductivity roughly in the range of 10^3 to 10^{-8} siemens per centimeter. Semiconductor materials are the foundation of modern electronics, including radio, computers, telephones, and many other devices.

Manufacturing

Manufacturing is the use of machines, tools and labor to produce goods for use or sale. The term may refer to a range of human activity, from handicraft to high tech, but is most commonly applied to industrial production, in which raw materials are transformed into finished goods on a large scale. Such finished goods may be used for manufacturing other, more complex products, such as aircraft, household appliances or automobiles, or sold to wholesalers, who in turn sell them to retailers, who then sell them to end users - the "consumers".

DC motor

A DC motor is an electric motor that runs on direct current (DC) electricity.

Brushed

The brushed DC electric motor generates torque directly from DC power supplied to the motor by using internal commutation, stationary permanent magnets, and rotating electrical magnets. Like all electric motors or generators, torque is produced by the principle of Lorentz force, which states that any current-carrying conductor placed within an external magnetic field experiences a torque or force known as Lorentz force.

Two-sided Laplace transform

In mathematics, the two-sided Laplace transform is an integral transform closely related to the Fourier transform, the Mellin transform, and the ordinary or one-sided Laplace transform. If $f(t)$ is a real or complex valued function of the real variable t defined for all real numbers, then the two-sided Laplace transform is defined by the integral

$$\mathcal{B}\{f(t)\} = F(s) = \int_{-\infty}^{\infty} e^{-st} f(t)\, dt.$$

The integral is most commonly understood as an improper integral, which converges if and only if each of the integrals

$$\int_0^{\infty} e^{-st} f(t)\, dt, \quad \int_{-\infty}^{0} e^{-st} f(t)\, dt$$

exists. There seems to be no generally accepted notation for the two-sided transform, the $\mathcal{B}$ used here recalls "bilateral".

Coefficient

In mathematics, a coefficient is a constant multiplicative factor of a specific object. For example, in the expression $9x^2$, the coefficient of x^2 is 9.

The object can be such things as a variable, a vector, a function, etc.

Inverse Laplace transform

In mathematics, the inverse Laplace transform of F(s) is the function f(t) which has the property

$$\mathcal{L}\{f(t)\} = F(s),$$

where $\mathcal{L}$ is the Laplace transform.

It can be proven, that if a function F(s) has the inverse Laplace transform f(t), i.e. f is a piecewise-continuous and exponentially-restricted real function f satisfying the condition

$$\mathcal{L}\{f(t)\} = F(s)$$

then f(t) is uniquely determined (considering functions which differ from each other only on a point set having Lebesgue measure zero as the same).

The Laplace transform and the inverse Laplace transform together have a number of properties that make them useful for analysing linear dynamic systems.

Final value theorem

In mathematical analysis, the final value theorem is one of several similar theorems used to relate frequency domain expressions to the time domain behavior as time approaches infinity. A final value theorem allows the time domain behavior to be directly calculated by taking a limit of a frequency domain expression, as opposed to converting to a time domain expression and taking its limit.

Mathematically, if

$$\lim_{t \to \infty} f(t)$$

exists, then

$$\lim_{t \to \infty} f(t) = \lim_{s \to 0} sF(s)$$

where F(s) is the Laplace transform of the signal f(t).

Simulink

Simulink, developed by MathWorks, is a commercial tool for modeling, simulating and analyzing multidomain dynamic systems. Its primary interface is a graphical block diagramming tool and a customizable set of block libraries. It offers tight integration with the rest of the MATLAB environment and can either drive MATLAB or be scripted from it.

Residue

In mathematics, more specifically complex analysis, the residue is a complex number equal to the contour integral of a meromorphic function along a path enclosing one of its singularities. Residues can be computed quite easily and, once known, allow the determination of general contour integrals via the residue theorem.

The residue of a meromorphic function f at an isolated singularity a, often denoted $\mathrm{Res}(f, a)$ is the unique value R such that f(z) − R / (z − a) has an analytic antiderivative in a punctured disk $0 < |z - a| < \delta$.

Differential

A differential is a device, usually but not necessarily employing gears, capable of transmitting torque and rotation through three shafts, almost always used in one of two ways: in one way, it receives one input and provides two outputs--this is found in most automobiles--and in the other way, it combines two inputs to create an output that is the sum, difference, or average, of the inputs.

In automobiles and other wheeled vehicles, the differential allows each of the driving roadwheels to rotate at different speeds, while for most vehicles supplying equal torque to each of them.

Purpose

A vehicle's wheels rotate at different speeds, mainly when turning corners.

Solution

In chemistry, a solution is a homogeneous mixture composed of two or more substances. In such a mixture, a solute is dissolved in another substance, known as a solvent. The solvent does the dissolving.

Poly

Poly can have multiple meanings:

· As a prefix, often meaning more than one or many (eg, Polyvalent - capable of many valences)

· As a feminine given name (Poly, also spelled Polly). Sometimes also used as a nickname for surnames such as 'Polyblank'

Poly is also commonly used as an abbreviation for various words and phrases including:

Science and technology

· Polyethylene, Polyurethane and Polyester, all forms of plastic Polymers

· Polysilicon, a state of silicon used as a gate material in MOSFET transistors and CMOS microchips

· Poly-A, a step in the transcription of DNA to RNA in which multiple adenosine nucleosides are added to the 3' end of the mRNA sequence.

· Polymorphism, Polymorphism describes multiple possible states for a single property

· Polyspheric, a shape composed of more than one sphere, a term used by Chrysler to identify their hemi-derived but lower cost V8 engines built 1955-67

· Polygraph, or 'lie detector' (particularly in Defense ' Intelligence fields)

· Poly, a 4 level computer game created by (in no particular order) Brian Giaime, Peter Haas, Tulio Correa, and Charles Engelhart, students at DigiPen Institute of Technology.In Poly, the player is a rolling figure attempting to collect coins and gems, while finishing levels quickly.

Mathematics and logic

· Polygon, especially in relation to 3D graphics as used in terms such as high-Poly

· Polynomial, a type of mathematical function

Other

· Polyamory, Polygamy, and certain other forms of nonmonogamy beginning in Poly

· Polyclinic, a clinic where a wide range of health care services (including diagnostics) can be obtained without the need for an overnight stay

· Polyrunning a person good at all track events

· Polytechnic educational institutions

· Polymath, a person well educated in a wide variety of subjects or fields

· Polydactyly, having extra digits in hands or feet

· Polymita, Business Process Management Suite

· Poly Technologies, Chinese company

· China Poly Group, Chinese business group

· Polynesian, as in 'Being a Poly' by Jonah Takalua

· Polyglot, a person who uses many languages

:'.

Transfer function

A transfer function is a mathematical representation, in terms of spatial or temporal frequency, of the relation between the input and output of a linear time-invariant system. With optical imaging devices, for example, it is the Fourier transform of the point spread function (hence a function of spatial frequency) i.e. the intensity distribution caused by a point object in the field of view.

Explanation

Transfer functions are commonly used in the analysis of systems such as single-input single-output filters, typically within the fields of signal processing, communication theory, and control theory.

Bode plot

A Bode plot is a graph of the transfer function of a linear, time-invariant system versus frequency, plotted with a log-frequency axis, to show the system's frequency response. It is usually a combination of a Bode magnitude plot, expressing the magnitude of the frequency response gain, and a Bode phase plot, expressing the frequency response phase shift.

Overview

Among his several important contributions to circuit theory and control theory, engineer Hendrik Wade Bode (1905-1982), while working at Bell Labs in the United States in the 1930s, devised a simple but accurate method for graphing gain and phase-shift plots.

System analysis

System analysis in the field of electrical engineering characterizes electrical systems and their properties. System Analysis can be used to represent almost anything from population growth to audio speakers, electrical engineers often use it because of its direct relevance to many areas of their discipline, most notably signal processing and communication systems.

Characterization of systems

A system is characterized by how it responds to input signals.

Compensation

In engineering, compensation is planning for side effects or other unintended issues in a design. The design of an invention can itself also be to compensate for some other existing issue or exception.

One example is in a voltage-controlled crystal oscillator (VCXO), which is normally affected not only by voltage, but to a lesser extent by temperature.

Block

A block in the context of extraction of petroleum, (and natural gas) signifies a sub-division of an underground reservoir for the purpose of administering a certain production allowable.

Block diagram

Block diagram is a diagram of a system, in which the principal parts or functions are represented by blocks connected by lines, that show the relationships of the blocks. They are heavily used in the engineering world in hardware design, software design, and process flow diagrams.

The block diagram is typically used for a higher level, less detailed description aimed more at understanding the overall concepts and less at understanding the details of implementation.

Negative feedback

Negative feedback occurs when the output of a system acts to oppose changes to the input of the system, with the result that the changes are attenuated. If the overall feedback of the system is negative, then the system will tend to be stable.

Overview

In many physical and biological systems, qualitatively different influences can oppose each other.

Block diagram

Block diagram is a diagram of a system, in which the principal parts or functions are represented by blocks connected by lines, that show the relationships of the blocks. They are heavily used in the engineering world in hardware design, software design, and process flow diagrams.

The block diagram is typically used for a higher level, less detailed description aimed more at understanding the overall concepts and less at understanding the details of implementation.

Reduction

In mathematics, reduction refers to the rewriting of an expression into a simpler form. For example, the process of rewriting a fraction into one with the smallest whole-number denominator possible (while keeping the numerator an integer) is called "reducing a fraction". Rewriting a radical (or "root") expression with the smallest possible whole number under the radical symbol is called "reducing a radical".

Series

A series is the sum of the terms of a sequence. Finite sequences and series have defined first and last terms, whereas infinite sequences and series continue indefinitely.

In mathematics, given an infinite sequence of numbers $\{ a_n \}$, a series is informally the result of adding all those terms together: $a_1 + a_2 + a_3 + \cdots$.

Stability

In probability theory, the stability of a random variable is the property that a linear combination of two independent copies of the variable has the same distribution, up to location and scale parameters. The distributions of random variables having this property are said to be "stable distributions". Results available in probability theory show that all possible distributions having this property are members of a four-parameter family of distributions.

Time constant

In physics and engineering, the Time constant, usually denoted by the Greek letter τ , is the risetime characterizing the response to a time-varying input of a first-order, linear time-invariant (LTI) system. The Time constant is the main characteristic unit of a first-order LTI system.

In the time domain, the usual choice to explore the time response is through the step response to a step input, or the impulse response to a Dirac delta function input.

Pyrometer

A Pyrometer is a non-contacting device that intercepts and measures thermal radiation, a process known as pyrometry. This device can be used to determine the temperature of an object's surface.

The word Pyrometer comes from the Greek word for fire, 'πυρ' , and meter, meaning to measure.

Pole

In the mathematical field of complex analysis, a pole of a meromorphic function is a certain type of singularity that behaves like the singularity of $\frac{1}{z^n}$ at z = 0. This means that, in particular, a pole of the function f(z) is a point a such that f(z) approaches infinity as z approaches a.

Formally, suppose U is an open subset of the complex plane C, a is an element of U and f : U {a} → C is a function which is holomorphic over its domain. If there exists a holomorphic function g : U → C and a positive integer n, such that for all z in U {a}

$$f(z) = \frac{g(z)}{(z-a)^n}$$

holds, then a is called a pole of f.

Damping

In physics, damping is any effect that tends to reduce the amplitude of oscillations in an oscillatory system, particularly the harmonic oscillator.

In mechanics, friction is one such damping effect. For many purposes the frictional force F_f can be modeled as being proportional to the velocity v of the object:

$$F_{\mathrm{f}} = -cv \,,$$

where c is the viscous damping coefficient, given in units of Newton-seconds per meter.

Nyquist plot

A Nyquist plot is used in automatic control and signal processing for assessing the stability of a system with feedback. It is represented by a graph in polar coordinates in which the gain and phase of a frequency response are plotted. The plot of these phasor quantities shows the phase as the angle and the magnitude as the distance from the origin.

Step response

The step response of a system in a given initial state consists of the time evolution of its outputs when its control inputs are Heaviside step functions. In electronic engineering and control theory, step response is the time behaviour of the outputs of a general system when its inputs change from zero to one in a very short time. The concept can be extended to the abstract mathematical notion of a dynamical system using an evolution parameter.

Rise time

In electronics, when describing a voltage or current step function, Rise time refers to the time required for a signal to change from a specified low value to a specified high value. Typically, these values are 10% and 90% of the step height. The output signal of a system is characterized also by fall time: both parameters depend on rise and fall times of input signal and on the characteristics of the system.

Specification

In regression analysis and related fields such as econometrics, specification is the process of converting a theory into a regression model. This process consists of selecting an appropriate functional form for the model and choosing which variables to include. Model specification is one of the first steps in regression analysis.

Chapter 3. Dynamic Response

Standard	A technical standard is an established norm or requirement. It is usually a formal document that establishes uniform engineering or technical criteria, methods, processes and practices. In contrast, a custom, convention, company product, corporate standard, etc which becomes generally accepted and dominant is often called a de facto standard.
Domain	In mathematical analysis, a domain is any connected open subset of a finite-dimensional vector space. This is a different concept than the domain of a function, though it is often used for that purpose, for example in partial differential equations and Sobolev spaces. Various degrees of smoothness of the boundary of the domain are required for various properties of functions defined on the domain to hold, such as integral theorems (Green's theorem, Stokes theorem), properties of Sobolev spaces, and to define measures on the boundary and spaces of traces (spaces of smooth functions defined on the boundary).
Overshoot	In microwave communication, overshoot refers to reception of microwave signals where they are not intended as a result of an unusual atmospheric, e.g., ionospheric, conditions.
Time domain	Time domain is a term used to describe the analysis of mathematical functions, or physical signals, with respect to time. In the Time domain, the signal or function`s value is known for all real numbers, for the case of continuous time, or at various separate instants in the case of discrete time. An oscilloscope is a tool commonly used to visualize real-world signals in the Time domain.
Step response	The step response of a system in a given initial state consists of the time evolution of its outputs when its control inputs are Heaviside step functions. In electronic engineering and control theory, step response is the time behaviour of the outputs of a general system when its inputs change from zero to one in a very short time. The concept can be extended to the abstract mathematical notion of a dynamical system using an evolution parameter.
Transfer function	A transfer function is a mathematical representation, in terms of spatial or temporal frequency, of the relation between the input and output of a linear time-invariant system. With optical imaging devices, for example, it is the Fourier transform of the point spread function (hence a function of spatial frequency) i.e. the intensity distribution caused by a point object in the field of view. Explanation

Transfer functions are commonly used in the analysis of systems such as single-input single-output filters, typically within the fields of signal processing, communication theory, and control theory.

Control system

A control system is a device or set of devices to manage, command, direct or regulate the behavior of other devices or systems.

There are two common classes of control systems, with many variations and combinations: logic or sequential controls, and feedback or linear controls. There is also fuzzy logic, which attempts to combine some of the design simplicity of logic with the utility of linear control.

Invariant

Invariant and invariance may have several meanings, among which are:

· invariant, an expression whose value doesn't change during program execution

· In computer science, a type in overriding that is neither covariant nor contravariant

· invariant something unaltered by a transformation, for example: taking a homotopy group functor on the category of topological spaces. Then, homomeomorphic spaces have isomorphic fundamental group.

BIBO stability

In electrical engineering, specifically signal processing and control theory, BIBO stability is a form of stability for linear signals and systems that take inputs. BIBO stands for Bounded-Input Bounded-Output. If a system is BIBO stable, then the output will be bounded for every input to the system that is bounded.

Capacitor

A capacitor is a passive electronic component consisting of a pair of conductors separated by a dielectric (insulator). When there is a potential difference (voltage) across the conductors, a static electric field develops in the dielectric that stores energy and produces a mechanical force between the conductors. An ideal capacitor is characterized by a single constant value, capacitance, measured in farads.

Parameter

In mathematics, statistics, and the mathematical sciences, a Parameter is a quantity that serves to relate functions and variables using a common variable (often t) when such a relationship would be difficult to explicate with an equation. In different contexts the term may have special uses.

· In a section on frequently misused words in his book The Writer's Art, James J. Kilpatrick quoted a letter from a correspondent, giving examples to illustrate the correct use of the word Parameter:

· A parametric equaliser is an audio filter that allows the frequency of maximum cut or boost to be set by one control, and the size of the cut or boost by another. These settings, the frequency level of the peak or trough, are two of the Parameters of a frequency response curve, and in a two-control equaliser they completely describe the curve.

Range

The maximal total range is the distance an aircraft can fly between takeoff and landing, as limited by fuel capacity in powered aircraft, or cross-country speed and environmental conditions in unpowered aircraft.

Ferry range means the maximum range the aircraft can fly. This usually means maximum fuel load, optionally with extra fuel tanks and minimum equipment.

Range

The maximal total range is the distance an aircraft can fly between takeoff and landing, as limited by fuel capacity in powered aircraft, or cross-country speed and environmental conditions in unpowered aircraft.

Ferry range means the maximum range the aircraft can fly. This usually means maximum fuel load, optionally with extra fuel tanks and minimum equipment.

Steady state	In chemistry, a steady state is a situation in which all state variables are constant in spite of ongoing processes that strive to change them. For an entire system to be at steady state, i.e. for all state variables of a system to be constant, there must be a flow through the system (compare mass balance). One of the most simple examples of such a system is the case of a bathtub with the tap open but without the bottom plug: after a certain time the water flows in and out at the same rate, so the water level (the state variable being Volume) stabilizes and the system is at steady state.
State	In functional analysis, a state on a C*-algebra is a positive linear functional of norm 1. The set of states of a C*-algebra A, sometimes denoted by S(A), is always a convex set. The extremal points of S(A) are called pure states. If A has a multiplicative identity, S(A) is compact in the weak*-topology.
System identification	In control engineering, the field of system identification uses statistical methods to build mathematical models of dynamical systems from measured data. System identification also includes the optimal design of experiments for efficiently generating informative data for fitting such models. Overview A dynamical mathematical model in this context is a mathematical description of the dynamic behavior of a system or process in either the time or frequency domain.
Linearization	In mathematics and its applications, linearization refers to finding the linear approximation to a function at a given point. In the study of dynamical systems, linearization is a method for assessing the local stability of an equilibrium point of a system of nonlinear differential equations or discrete dynamical systems. This method is used in fields such as engineering, physics, economics, and ecology.
Motion	In physics, motion is a change in position of an object with respect to time. Change in action is the result of an unapplied force. Motion is typically described in terms of velocity, acceleration, displacement, and time.

Fourier

Fourier most commonly refers to Joseph Fourier French mathematician and physicist, or the mathematics, physics, and engineering terms named in his honor for his work on the concepts underlying them:

- Mathematics tools:
  - The Fourier series, a weighted sum of sinusoids having a common period, the result of Fourier analysis of a periodic function
  - Fourier analysis, the description of functions as sums of sinusoids
  - The Fourier transform, the type of linear canonical transform that is the generalization of the Fourier series
  - The Fourier operator, the kernel of the Fredholm integral of the first kind that defines the continuous Fourier transform
  - Fourier inversion theorem, any one of several theorems by which Fourier inversioı recovers a function from its Fourier transform
  - List of Fourier-related transforms, a list of linear transformations of functions related to Fourier analysis
  - Short-time Fourier transform or short-term Fourier transform (STFT), a Fourier transform during a short term of time, used in the area of signal analysis
  - Fractional Fourier transform (FRFT), a linear transformation generalizing the Fourier transform, used in the area of harmonic analysis
  - Discrete-time Fourier transform (DTFT), the reverse of the Fourier series, a speciέ case of the Z-transform around the unit circle in the complex plane
  - Discrete Fourier transform (DFT), occasionally called the finite Fourier transform, the Fourier transform of a discrete periodic sequence (yielding discrete periodic frequencies), which can also be thought of as the DTFT of a finite-length sequenc evaluated at discrete frequencies
  - Fast Fourier transform (FFT), a fast algorithm for computing a Discrete Fourier transform
  - Generalized Fourier series, generalizations of Fourier series that are special case of decompositions over an orthonormal basis of an inner product space
- Physics and engineering:
  - The Fourier number (Fo) (also known as the Fourier modulus), a ratio $\alpha t / d^2$ of th rate of heat conduction αt to the rate of thermal energy storage d^2
  - Fourier transform spectroscopy, a measurement technique whereby spectra are collected based on measurements of the temporal coherence of a radiative source, using time-domain measurements of the electromagnetic radiation or othe type of radiation, including several methods such as the continuous wave Michelson or Fourier transform spectrometer and the pulsed Fourier transform spectrograph

People named Fourier

- Joseph Fourier French mathematician and physicist

Tape drive	A tape drive is a data storage device that reads and writes data on a magnetic tape. It is typically used for offline, archival data storage. Tape media generally has a favorable unit cost and long archival stability.
Rolling	Rolling is a combination of rotation (of a radially symmetric object) and translation of that object with respect to a surface (either one or the other moves), such that the two are in contact with each other without sliding. This is achieved by a rotational speed at the cylinder or circle of contact which is equal to the translational speed. Rolling of a round object typically requires less energy than sliding, therefore such an object will more easily move, if it experiences a force with a component along the surface, for instance gravity on a tilted surface; wind; pushing; pulling; an engine.
Magnetic levitation	Magnetic levitation, maglev, or magnetic suspension is a method by which an object is suspended with no support other than magnetic fields. Magnetic pressure is used to counteract the effects of the gravitational and any other accelerations. Earnshaw's theorem proves that using only static ferromagnetism it is impossible to stably levitate against gravity, but servomechanisms, the use of diamagnetic materials, superconduction, or systems involving eddy currents permit this to occur.
Levitation	Levitation is the process by which an object is suspended by a physical force against gravity, in a stable position without solid physical contact. A number of different techniques have been developed to levitate matter, including the aerodynamic, magnetic, acoustic, electromagnetic, electrostatic, gas film, and optical levitation methods. Physics For levitation on Earth, first, a force is required directed vertically upwards and equal to the gravitational force, second, for any small displacement of the levitating object, a returning force should appear to stabilize it.
Electric power	Electric power is the rate at which electrical energy is transferred by an electric circuit. The SI unit of power is the watt.

When electric current flows in a circuit, it can transfer energy to do mechanical or thermodynamic work.

Power

In physics, power is the rate at which work is performed or energy is converted

If ΔW is the amount of work performed during a period of time of duration Δt, the average power P_{avg} over that period is given by the formula

$$P_{\text{avg}} = \frac{\Delta W}{\Delta t}.$$

It is the average amount of work done or energy converted per unit of time. The average power is often simply called "power" when the context makes it clear.

The instantaneous power is then the limiting value of the average power as the time interval Δt approaches zero.

Perspective	Perspective, in context of vision and visual perception, is the way in which objects appear to the eye based on their spatial attributes; or their dimensions and the position of the eye relative to the objects. There are two main meanings of the term: linear perspective and aerial perspective. Linear perspective As objects become more distant they appear smaller because their visual angle decreases.
Reference	Hydrogen silsesquioxane is a negative tone electron-beam resist. Thin HSQ layers have been used to show very high resolution (~10 nm) feature sizes. Reference · http://cat.inist.fr/?aModele=afficheN'cpsidt=18807320 :'.
Sensitivity	Sensitivity of explosives is the degree to which an explosive can be initiated by impact, heat, or friction. Sensitivity, along with stability and brisance are three of the most significant properties of explosives that affect their use and application. All explosive compounds have a certain amount of energy required to initiate.

Tracking

In particle physics, the tracking is the act of measuring the direction and magnitude of charged particles momenta. The particles entering a tracker (the device used for tracking), release part of their energy in the device: the tracker has to be finely segmented in order to be able to reconstruct with good precision where the particle passed. Since the tracking is usually made in a region where a magnetic field is present, it is possible to reconstruct part of the helix made by the particle inside the tracker (that is called track), and from the track parameters, and by knowing the mass of the particle under study (which is known by the use of particle identification), it is possible to reconstruct the actual direction and magnitude of the particle momenta.

Stability

In probability theory, the stability of a random variable is the property that a linear combination of two independent copies of the variable has the same distribution, up to location and scale parameters. The distributions of random variables having this property are said to be "stable distributions". Results available in probability theory show that all possible distributions having this property are members of a four-parameter family of distributions.

Control system

A control system is a device or set of devices to manage, command, direct or regulate the behavior of other devices or systems.

There are two common classes of control systems, with many variations and combinations: logic or sequential controls, and feedback or linear controls. There is also fuzzy logic, which attempts to combine some of the design simplicity of logic with the utility of linear control.

Integral

Integration is an important concept in mathematics and, together with differentiation, is one of the two main operations in calculus. Given a function f of a real variable x and an interval [a, b] of the real line, the definite integral

is defined informally to be the net signed area of the region in the xy-plane bounded by the graph of f, the x-axis, and the vertical lines x = a and x = b.

The principles of integration were formulated independently by Isaac Newton and Gottfried Leibniz in the late 17th century.

Property

In modern philosophy, mathematics, and logic, a property is an attribute of an object; a red object is said to have the property of redness. The property may be considered a form of object in its own right, able to possess other properties. It differs from the logical concept of class by not having any concept of extensionality, and from the philosophical concept of class in that a property is considered to be distinct from the objects which possess it.

Regulator

In automatic control, a regulator is a device which has the function of maintaining a designated characteristic. It performs the activity of managing or maintaining a range of values in a machine. The measurable property of a device is managed closely by specified conditions or an advance set value; or it can be a variable according to a predetermined arrangement scheme.

Speed

In kinematics, the speed of an object is the magnitude of its velocity (the rate of change of its position); it is thus a scalar quantity. The average speed of an object in an interval of time is the distance traveled by the object divided by the duration of the interval; the instantaneous speed is the limit of the average speed as the duration of the time interval approaches zero .

Like velocity, speed has the dimensions of a length divided by a time; the SI unit of speed is the meter per second, but the most usual unit of speed in everyday usage is the kilometer per hour or, in the USA and the UK, miles per hour.

Tachometer

A tachometer is an instrument that measures the rotation speed of a shaft or disk, as in a motor or other machine. The device usually displays the revolutions per minute (RPM) on a calibrated analogue dial, but digital displays are increasingly common. The term comes from Greek Ταχος, tachos, "speed", and metron, "to measure".

DC motor

A DC motor is an electric motor that runs on direct current (DC) electricity.

Brushed

The brushed DC electric motor generates torque directly from DC power supplied to the motor by using internal commutation, stationary permanent magnets, and rotating electrical magnets. Like all electric motors or generators, torque is produced by the principle of Lorentz force, which states that any current-carrying conductor placed within an external magnetic field experiences a torque or force known as Lorentz force.

PID controller

A proportional-integral-derivative controller (PID controller) is a generic control loop feedback mechanism (controller) widely used in industrial control systems - a PID is the most commonly used feedback controller. A PID controller calculates an "error" value as the difference between a measured process variable and a desired setpoint. The controller attempts to minimize the error by adjusting the process control inputs.

Controller

In control theory, a controller is a device which monitors and affects the operational conditions of a given dynamical system. The operational conditions are typically referred to as output variables of the system which can be affected by adjusting certain input variables. For example, the heating system of a house can be equipped with a thermostat (controller) for sensing air temperature (output variable) which can turn on or off a furnace or heater when the air temperature becomes too low or too high.

Digital

A digital system is a data technology that uses discrete (discontinuous) values. By contrast, non-digital systems use a continuous range of values to represent information. Although digital representations are discrete, the information represented can be either discrete, such as numbers, letters or icons, or continuous, such as sounds, images, and other measurements of continuous systems.

Curve

In mathematics, a Curve consists of the points through which a continuously moving point passes. This notion captures the intuitive idea of a geometrical one-dimensional object, which furthermore is connected in the sense of having no discontinuities or gaps. Simple examples include the sine wave as the basic Curve underlying simple harmonic motion, and the parabola.

Reaction

The third of Newton's laws of motion of classical mechanics states that forces always occur in pairs. Every action is accompanied by a reaction of equal magnitude but opposite direction. This principle is commonly known in the Latin language as actio et reactio.

Period	In the periodic table of the elements, elements are arranged in a series of rows (or periods) so that those with similar properties appear in vertical columns. Elements of the same period have the same number of electron shells; with each group across a period, the elements have one more proton and electron and become less metallic. This arrangement reflects the periodic recurrence of similar properties as the atomic number increases.
Ratio	In mathematics, a ratio expresses the magnitude of quantities relative to each other. Specifically, the ratio of two quantities indicates how many times the first quantity is contained in the second and may be expressed algebraically as their quotient. Mathematically, a proportion is defined as the equality of two ratios.
Heat exchanger	A heat exchanger is a device built for efficient heat transfer from one medium to another. The media may be separated by a solid wall, so that they never mix, or they may be in direct contact. They are widely used in space heating, refrigeration, air conditioning, power plants, chemical plants, petrochemical plants, petroleum refineries, natural gas processing, and sewage treatment.
Signal	In the fields of communications, signal processing, and in electrical engineering more generally, a signal is any time-varying or spatial-varying quantity. In the physical world, any quantity measurable through time or over space can be taken as a signal. Within a complex society, any set of human information or machine data can also be taken as a signal.

Design methods	Design Methods is a broad area that focuses on: • Divergence - Exploring possibilities and constraints of inherited situations by applying critical thinking through qualitative and quantitative research methods to create new understanding (problem space) toward better design solutions • Transformation - Redefining specifications of design solutions which can lead to better guidelines for traditional and contemporary design activities (architecture, graphic, industrial, information, interaction, et al). and/or multidisciplinary response • Convergence - Prototyping possible scenarios for better design solutions that incrementally or significantly improve the originally inherited situation • Sustainability - Managing the process of exploring, redefining and prototyping of design solutions continually over time • Articulation - the visual relationship between the parts and the whole. The goal of design methods is to gain key insights or unique essential truths resulting in more holistic solutions in order to achieve better experiences for users with products, services, environments and systems they rely upon. Insight, in this case, is clear and deep investigation of a situation through design methods, thereby grasping the inner nature of things intuitively.
Perspective	Perspective, in context of vision and visual perception, is the way in which objects appear to the eye based on their spatial attributes; or their dimensions and the position of the eye relative to the objects. There are two main meanings of the term: linear perspective and aerial perspective. Linear perspective As objects become more distant they appear smaller because their visual angle decreases.
Break-in	Break-in, also known as run-in or running in, is the procedure of conditioning a new piece of equipment by giving it an initial period of running, usually under light load, but sometimes under heavy load or normal load. It is generally a process of moving parts wearing against each other to produce the last small bit of size and shape adjustment that will settle them into a stable relationship for the rest of their working life.

One of the commonest examples of break-in is engine break-in for petrol engines and diesel engines.

Rise time

In electronics, when describing a voltage or current step function, Rise time refers to the time required for a signal to change from a specified low value to a specified high value. Typically, these values are 10% and 90% of the step height. The output signal of a system is characterized also by fall time: both parameters depend on rise and fall times of input signal and on the characteristics of the system.

Positive

A positive is a film or paper record of a scene that represents the color and luminance of objects in that scene with the same colors and luminances (as near as the medium will allow). Color transparencies are an example of positive photography: the range of colors presented in the medium is limited by the tonal range of the original image (dark and light areas correspond). It is opposed to a negative where colors and luminances are reversed: this is due to the chemical or electrical processes involved in recording the scene.

Asymptote

In analytic geometry, an asymptote of a curve is a line such that the distance between the curve and the line approaches zero as they tend to infinity. Some sources include the requirement that the curve may not cross the line infinitely often, but this is unusual for modern authors. In some contexts, such as algebraic geometry, an asymptote is defined as a line which is tangent to a curve at infinity.

Angle

In geometry, an angle is the figure formed by two rays sharing a common endpoint, called the vertex of the angle. The magnitude of the angle is the "amount of rotation" that separates the two rays, and can be measured by considering the length of circular arc swept out when one ray is rotated about the vertex to coincide with the other . Where there is no possibility of confusion, the term "angle" is used interchangeably for both the geometric configuration itself and for its angular magnitude (which is simply a numerical quantity).

Transfer function

A transfer function is a mathematical representation, in terms of spatial or temporal frequency, of the relation between the input and output of a linear time-invariant system. With optical imaging devices, for example, it is the Fourier transform of the point spread function (hence a function of spatial frequency) i.e. the intensity distribution caused by a point object in the field of view.

Explanation

Transfer functions are commonly used in the analysis of systems such as single-input single-output filters, typically within the fields of signal processing, communication theory, and control theory.

Pole

In the mathematical field of complex analysis, a pole of a meromorphic function is a certain type of singularity that behaves like the singularity of $\frac{1}{z^n}$ at z = 0. This means that, in particular, a pole of the function f(z) is a point a such that f(z) approaches infinity as z approaches a.

Formally, suppose U is an open subset of the complex plane C, a is an element of U and f : U {a} → C is a function which is holomorphic over its domain. If there exists a holomorphic function g : U → C and a positive integer n, such that for all z in U {a}

$$f(z) = \frac{g(z)}{(z-a)^n}$$

holds, then a is called a pole of f.

Transition

In music, the transition is the middle section or formal function, while the main theme is the beginning, and the subordinate theme is the ending. It may traditionally be a part of the Sonata form's exposition in which the composer modulates from the key of the first subject to the key of the second, though many Classical era works move straight from first to second subject groups without any transition.

For example, transition may be defined as different from a subordinate theme (rondo form) or a developmental core.

Simulink

Simulink, developed by MathWorks, is a commercial tool for modeling, simulating and analyzing multidomain dynamic systems. Its primary interface is a graphical block diagramming tool and a customizable set of block libraries. It offers tight integration with the rest of the MATLAB environment and can either drive MATLAB or be scripted from it.

Pole

In the mathematical field of complex analysis, a pole of a meromorphic function is a certain type of singularity that behaves like the singularity of $\frac{1}{z^n}$ at z = 0. This means that, in particular, a pole of the function f(z) is a point a such that f(z) approaches infinity as z approaches a.

Formally, suppose U is an open subset of the complex plane C, a is an element of U and f : U {a} → C is a function which is holomorphic over its domain. If there exists a holomorphic function g : U → C and a positive integer n, such that for all z in U {a}

$$f(z) = \frac{g(z)}{(z-a)^n}$$

holds, then a is called a pole of f.

Case

A case of some merchandise is a collection of items packaged together. In the United States, typically a standard case contains a certain number of items depending on what the merchandise is. For consumer foodstuffs such as canned goods, soda, cereal and such, a case is typically 24 items, however cases may range from 12 to 36, typically in multiples of six.

Compensation

In engineering, compensation is planning for side effects or other unintended issues in a design. The design of an invention can itself also be to compensate for some other existing issue or exception.

One example is in a voltage-controlled crystal oscillator (VCXO), which is normally affected not only by voltage, but to a lesser extent by temperature.

DC motor	A DC motor is an electric motor that runs on direct current (DC) electricity. Brushed The brushed DC electric motor generates torque directly from DC power supplied to the motor by using internal commutation, stationary permanent magnets, and rotating electrical magnets. Like all electric motors or generators, torque is produced by the principle of Lorentz force, which states that any current-carrying conductor placed within an external magnetic field experiences a torque or force known as Lorentz force.
Dynamics	In the field of physics, the study of the causes of motion and changes in motion is dynamics. In other words the study of forces and why objects are in motion. Dynamics includes the study of the effect of torques on motion.
Estimator	In statistics, an estimator is a rule for calculating an estimate of a given quantity based on observed data: thus the rule and its result (the estimate) are distinguished This is in contrast to an interval estimator, where the result would be a range of plausible values (or vectors or functions).
Compensation	In engineering, compensation is planning for side effects or other unintended issues in a design. The design of an invention can itself also be to compensate for some other existing issue or exception. One example is in a voltage-controlled crystal oscillator (VCXO), which is normally affected not only by voltage, but to a lesser extent by temperature.
Implementation	Implementation is the realization of an application, or execution of a plan, idea, model, design, specification, standard, algorithm, or policy. Computer Science

In computer science, an implementation is a realization of a technical specification or algorithm as a program, software component, or other computer system through programming and deployment. Many implementations may exist for a given specification or standard.

Step response

The step response of a system in a given initial state consists of the time evolution of its outputs when its control inputs are Heaviside step functions. In electronic engineering and control theory, step response is the time behaviour of the outputs of a general system when its inputs change from zero to one in a very short time. The concept can be extended to the abstract mathematical notion of a dynamical system using an evolution parameter.

Digital

A digital system is a data technology that uses discrete (discontinuous) values. By contrast, non-digital systems use a continuous range of values to represent information. Although digital representations are discrete, the information represented can be either discrete, such as numbers, letters or icons, or continuous, such as sounds, images, and other measurements of continuous systems.

Implementation

Implementation is the realization of an application, or execution of a plan, idea, model, design, specification, standard, algorithm, or policy.

Computer Science

In computer science, an implementation is a realization of a technical specification or algorithm as a program, software component, or other computer system through programming and deployment. Many implementations may exist for a given specification or standard.

Lead

Lead refers to which set of legs, left or right, leads or advances forward to a greater extent when a quadruped animal is cantering, galloping, or leaping. The feet on the leading side touch the ground forward of its partner. On the "left lead", the animal's left legs lead.

Controller

In control theory, a controller is a device which monitors and affects the operational conditions of a given dynamical system. The operational conditions are typically referred to as output variables of the system which can be affected by adjusting certain input variables. For example, the heating system of a house can be equipped with a thermostat (controller) for sensing air temperature (output variable) which can turn on or off a furnace or heater when the air temperature becomes too low or too high.

Continuous	In probability theory, a probability distribution is called continuous if its cumulative distribution function is continuous. This is equivalent to saying that for random variables X with the distribution in question, Pr[X = a] = 0 for all real numbers a, i.e.: the probability that X attains the value a is zero, for any number a. If the distribution of X is continuous then X is called a continuous random variable.
Discrete system	A discrete system is a system with a countable number of states. Discrete systems may be contrasted with continuous systems, which may also be called analog systems. A final discrete system is often modeled with a directed graph and is analyzed for correctness and complexity according to computational theory.
Autopilot	An autopilot is a mechanical, electrical, or hydraulic system used to guide a vehicle without assistance from a human being. An autopilot can refer specifically to aircraft, self-steering gear for boats, or auto guidance of space craft and missiles. The autopilot of an aircraft is sometimes referred to as "George".
Block	A block in the context of extraction of petroleum, (and natural gas) signifies a sub-division of an underground reservoir for the purpose of administering a certain production allowable.
Block diagram	Block diagram is a diagram of a system, in which the principal parts or functions are represented by blocks connected by lines, that show the relationships of the blocks. They are heavily used in the engineering world in hardware design, software design, and process flow diagrams. The block diagram is typically used for a higher level, less detailed description aimed more at understanding the overall concepts and less at understanding the details of implementation.
Parameter	In mathematics, statistics, and the mathematical sciences, a Parameter is a quantity that serves to relate functions and variables using a common variable (often t) when such a relationship would be difficult to explicate with an equation. In different contexts the term may have special uses. · In a section on frequently misused words in his book The Writer's Art, James J. Kilpatrick quoted a letter from a correspondent, giving examples to illustrate the correct use of the word Parameter:

· A parametric equaliser is an audio filter that allows the frequency of maximum cut or boost to be set by one control, and the size of the cut or boost by another. These settings, the frequency level of the peak or trough, are two of the Parameters of a frequency response curve, and in a two-control equaliser they completely describe the curve.

Lyapunov stability

In mathematics, the notion of Lyapunov stability occurs in the study of dynamical systems. In simple terms, if all solutions of the dynamical system that start out near an equilibrium point x_e stay near x_e forever, then x_e is Lyapunov stable. More strongly, if x_e is Lyapunov stable and all solutions that start out near x_e converge to x_e, then x_e is asymptotically stable.

Stability

In probability theory, the stability of a random variable is the property that a linear combination of two independent copies of the variable has the same distribution, up to location and scale parameters. The distributions of random variables having this property are said to be "stable distributions". Results available in probability theory show that all possible distributions having this property are members of a four-parameter family of distributions.

Heat exchanger

A heat exchanger is a device built for efficient heat transfer from one medium to another. The media may be separated by a solid wall, so that they never mix, or they may be in direct contact. They are widely used in space heating, refrigeration, air conditioning, power plants, chemical plants, petrochemical plants, petroleum refineries, natural gas processing, and sewage treatment.

Design methods	Design Methods is a broad area that focuses on: • Divergence - Exploring possibilities and constraints of inherited situations by applying critical thinking through qualitative and quantitative research methods to create new understanding (problem space) toward better design solutions • Transformation - Redefining specifications of design solutions which can lead to better guidelines for traditional and contemporary design activities (architecture, graphic, industrial, information, interaction, et al). and/or multidisciplinary response • Convergence - Prototyping possible scenarios for better design solutions that incrementally or significantly improve the originally inherited situation • Sustainability - Managing the process of exploring, redefining and prototyping of design solutions continually over time • Articulation - the visual relationship between the parts and the whole. The goal of design methods is to gain key insights or unique essential truths resulting in more holistic solutions in order to achieve better experiences for users with products, services, environments and systems they rely upon. Insight, in this case, is clear and deep investigation of a situation through design methods, thereby grasping the inner nature of things intuitively.
Perspective	Perspective, in context of vision and visual perception, is the way in which objects appear to the eye based on their spatial attributes; or their dimensions and the position of the eye relative to the objects. There are two main meanings of the term: linear perspective and aerial perspective. Linear perspective As objects become more distant they appear smaller because their visual angle decreases.
Stability	In probability theory, the stability of a random variable is the property that a linear combination of two independent copies of the variable has the same distribution, up to location and scale parameters. The distributions of random variables having this property are said to be "stable distributions". Results available in probability theory show that all possible distributions having this property are members of a four-parameter family of distributions.

Chapter 6. The Frequency-Response Design Method

Frequency	Frequency is the number of occurrences of a repeating event per unit time. It is also referred to as temporal frequency. The period is the duration of one cycle in a repeating event, so the period is the reciprocal of the frequency.
Frequency response	Frequency response is the measure of any system's output spectrum in response to an input signal. In the audible range it is usually referred to in connection with electronic amplifiers, microphones and loudspeakers. Radio spectrum frequency response can refer to measurements of coaxial cables, category cables, video switchers and wireless communications devices.
Control system	A control system is a device or set of devices to manage, command, direct or regulate the behavior of other devices or systems. There are two common classes of control systems, with many variations and combinations: logic or sequential controls, and feedback or linear controls. There is also fuzzy logic, which attempts to combine some of the design simplicity of logic with the utility of linear control.
Capacitor	A capacitor is a passive electronic component consisting of a pair of conductors separated by a dielectric (insulator). When there is a potential difference (voltage) across the conductors, a static electric field develops in the dielectric that stores energy and produces a mechanical force between the conductors. An ideal capacitor is characterized by a single constant value, capacitance, measured in farads.
Semiconductor	A semiconductor is a material with electrical conductivity due to electron flow (as opposed to ionic conductivity) intermediate in magnitude between that of a conductor and an insulator. This means a conductivity roughly in the range of 10^3 to 10^{-8} siemens per centimeter. Semiconductor materials are the foundation of modern electronics, including radio, computers, telephones, and many other devices.
Manufacturing	Manufacturing is the use of machines, tools and labor to produce goods for use or sale. The term may refer to a range of human activity, from handicraft to high tech, but is most commonly applied to industrial production, in which raw materials are transformed into finished goods on a large scale. Such finished goods may be used for manufacturing other, more complex products, such as aircraft, household appliances or automobiles, or sold to wholesalers, who in turn sell them to retailers, who then sell them to end users - the "consumers".

Bandwidth	Bandwidth is the difference between the upper and lower frequencies in a contiguous set of frequencies. It is typically measured in hertz, and may sometimes refer to passband bandwidth, sometimes to baseband bandwidth, depending on context. Passband bandwidth is the difference between the upper and lower cutoff frequencies of, for example, an electronic filter, a communication channel, or a signal spectrum.
Transfer function	A transfer function is a mathematical representation, in terms of spatial or temporal frequency, of the relation between the input and output of a linear time-invariant system. With optical imaging devices, for example, it is the Fourier transform of the point spread function (hence a function of spatial frequency) i.e. the intensity distribution caused by a point object in the field of view. Explanation Transfer functions are commonly used in the analysis of systems such as single-input single-output filters, typically within the fields of signal processing, communication theory, and control theory.
Bode plot	A Bode plot is a graph of the transfer function of a linear, time-invariant system versus frequency, plotted with a log-frequency axis, to show the system's frequency response. It is usually a combination of a Bode magnitude plot, expressing the magnitude of the frequency response gain, and a Bode phase plot, expressing the frequency response phase shift. Overview Among his several important contributions to circuit theory and control theory, engineer Hendrik Wade Bode (1905-1982), while working at Bell Labs in the United States in the 1930s, devised a simple but accurate method for graphing gain and phase-shift plots.

Decade

One decade is a factor of 10 difference between two numbers (an order of magnitude difference) measured on a logarithmic scale. It is especially useful when referring to frequencies and when describing frequency response of electronic systems, such as audio amplifiers and filters.

Calculations

The factor-of-ten in a decade can be in either direction: so one decade up from 100 Hz is 1000 Hz, and one decade down is 10 Hz.

Decibel

The decibel is a logarithmic unit that indicates the ratio of a physical quantity (usually power or intensity) relative to a specified or implied reference level. A ratio in decibels is ten times the logarithm to base 10 of the ratio of two power quantities. Being a ratio of two measurements of a physical quantity in the same units, it is a dimensionless unit.

Pole

In the mathematical field of complex analysis, a pole of a meromorphic function is a certain type of singularity that behaves like the singularity of $\frac{1}{z^n}$ at z = 0. This means that, in particular, a pole of the function f(z) is a point a such that f(z) approaches infinity as z approaches a.

Formally, suppose U is an open subset of the complex plane C, a is an element of U and f : U {a} → C is a function which is holomorphic over its domain. If there exists a holomorphic function g : U → C and a positive integer n, such that for all z in U {a}

$$f(z) = \frac{g(z)}{(z-a)^n}$$

holds, then a is called a pole of f.

Asymptote	In analytic geometry, an asymptote of a curve is a line such that the distance between the curve and the line approaches zero as they tend to infinity. Some sources include the requirement that the curve may not cross the line infinitely often, but this is unusual for modern authors. In some contexts, such as algebraic geometry, an asymptote is defined as a line which is tangent to a curve at infinity.
Angle	In geometry, an angle is the figure formed by two rays sharing a common endpoint, called the vertex of the angle. The magnitude of the angle is the "amount of rotation" that separates the two rays, and can be measured by considering the length of circular arc swept out when one ray is rotated about the vertex to coincide with the other . Where there is no possibility of confusion, the term "angle" is used interchangeably for both the geometric configuration itself and for its angular magnitude (which is simply a numerical quantity).
Phase margin	In electronic amplifiers, phase margin is the difference between the phase, measured in degrees, of an amplifier's output signal (relative to its input) and 180°, as a function of frequency. The Phase margin is taken as positive at frequencies below where the open-loop phase first crosses 180°, i.e. the signal becomes inverted, or antiphase; it is negative beyond. In negative feedback, a negative Phase margin at a frequency where the loop gain exceeds unity guarantees instability, thus positive Phase margin is a "safety margin" that ensures proper operation of an amplifier and, more generally, active filters, under various loads.
Curve	In mathematics, a Curve consists of the points through which a continuously moving point passes. This notion captures the intuitive idea of a geometrical one-dimensional object, which furthermore is connected in the sense of having no discontinuities or gaps. Simple examples include the sine wave as the basic Curve underlying simple harmonic motion, and the parabola.
Appendage	In invertebrate biology, an appendage is an external body part, or natural prolongation, that protrudes from an organism's body, such as a vertebrate's limbs. It is a general term that covers any of the homologous body parts that may extend from a body segment. These include antennae, mouthparts (including mandibles, maxillae and maxillipeds), wings, elytra, gills, walking legs (pereiopods), swimming legs (pleopods), sexual organs (gonopods), and parts of the tail (uropods).
Coefficient	In mathematics, a coefficient is a constant multiplicative factor of a specific object. For example, in the expression $9x^2$, the coefficient of x^2 is 9. The object can be such things as a variable, a vector, a function, etc.

Nyquist stability criterion

The Nyquist stability criterion provides a simple test for stability of a closed-loop control system by examining the open-loop system's Nyquist plot. Under many circumstances, stability of the closed-loop control system may be determined directly by computing the poles of the closed-loop transfer function. In contrast, the Nyquist stability criterion allows stability to be determined without computing the closed-loop poles.

Argument

In mathematics, arg is a function operating on complex numbers (visualised as a flat plane). It gives the angle between the line joining the point to the origin and the positive real axis, shown as φ in figure 1 opposite, known as an argument of the point (that is, the angle between the half-lines of the position vector representing the number and the positive real axis).

Arguments are defined in two equivalent ways:

- Geometrically, in relation to an Argand diagram, arg z is the angle φ from the positive real axis to the vector representing z.

Argument principle

In complex analysis, the Argument principle states that if f(z) is a meromorphic function inside and on some closed contour C, with f having no zeros or poles on C, then the following formula holds

$$\oint_C \frac{f'(z)}{f(z)}\,dz = 2\pi i(N - P)$$

where N and P denote respectively the number of zeros and poles of f(z) inside the contour C, with each zero and pole counted as many times as its multiplicity and order respectively. This statement of the theorem assumes that the contour C is simple, that is, without self-intersections, and that it is oriented counter-clockwise.

More generally, suppose that C is a curve, oriented counter-clockwise, which is contractible to a point inside an open set Ω in the complex plane.

Nyquist plot

A Nyquist plot is used in automatic control and signal processing for assessing the stability of a system with feedback. It is represented by a graph in polar coordinates in which the gain and phase of a frequency response are plotted. The plot of these phasor quantities shows the phase as the angle and the magnitude as the distance from the origin.

Compensation

In engineering, compensation is planning for side effects or other unintended issues in a design. The design of an invention can itself also be to compensate for some other existing issue or exception.

One example is in a voltage-controlled crystal oscillator (VCXO), which is normally affected not only by voltage, but to a lesser extent by temperature.

DC motor

A DC motor is an electric motor that runs on direct current (DC) electricity.

Brushed

The brushed DC electric motor generates torque directly from DC power supplied to the motor by using internal commutation, stationary permanent magnets, and rotating electrical magnets. Like all electric motors or generators, torque is produced by the principle of Lorentz force, which states that any current-carrying conductor placed within an external magnetic field experiences a torque or force known as Lorentz force.

Damping

In physics, damping is any effect that tends to reduce the amplitude of oscillations in an oscillatory system, particularly the harmonic oscillator.

In mechanics, friction is one such damping effect. For many purposes the frictional force F_f can be modeled as being proportional to the velocity v of the object:

$$F_f = -cv ,$$

where c is the viscous damping coefficient, given in units of Newton-seconds per meter.

Overshoot

In microwave communication, overshoot refers to reception of microwave signals where they are not intended as a result of an unusual atmospheric, e.g., ionospheric, conditions.

Ratio

In mathematics, a ratio expresses the magnitude of quantities relative to each other. Specifically, the ratio of two quantities indicates how many times the first quantity is contained in the second and may be expressed algebraically as their quotient. Mathematically, a proportion is defined as the equality of two ratios.

Block

A block in the context of extraction of petroleum, (and natural gas) signifies a sub-division of an underground reservoir for the purpose of administering a certain production allowable.

Block diagram

Block diagram is a diagram of a system, in which the principal parts or functions are represented by blocks connected by lines, that show the relationships of the blocks. They are heavily used in the engineering world in hardware design, software design, and process flow diagrams.

The block diagram is typically used for a higher level, less detailed description aimed more at understanding the overall concepts and less at understanding the details of implementation.

Estimator

In statistics, an estimator is a rule for calculating an estimate of a given quantity based on observed data: thus the rule and its result (the estimate) are distinguished This is in contrast to an interval estimator, where the result would be a range of plausible values (or vectors or functions).

Derivative

In chemistry, a derivative is a compound that is derived from a similar compound by some chemical or physical process. In the past it was also used to mean a compound that can be imagined to arise from another compound, if one atom is replaced with another atom or group of atoms, but modern chemical language now uses the term structural analogue for this meaning - thus eliminating ambiguity of both terms. The term "structural analogue" is common in organic chemistry.

Simulink

Simulink, developed by MathWorks, is a commercial tool for modeling, simulating and analyzing multidomain dynamic systems. Its primary interface is a graphical block diagramming tool and a customizable set of block libraries. It offers tight integration with the rest of the MATLAB environment and can either drive MATLAB or be scripted from it.

Controller

In control theory, a controller is a device which monitors and affects the operational conditions of a given dynamical system. The operational conditions are typically referred to as output variables of the system which can be affected by adjusting certain input variables. For example, the heating system of a house can be equipped with a thermostat (controller) for sensing air temperature (output variable) which can turn on or off a furnace or heater when the air temperature becomes too low or too high.

Continuous

In probability theory, a probability distribution is called continuous if its cumulative distribution function is continuous. This is equivalent to saying that for random variables X with the distribution in question, Pr[X = a] = 0 for all real numbers a, i.e.: the probability that X attains the value a is zero, for any number a. If the distribution of X is continuous then X is called a continuous random variable.

Parameter

In mathematics, statistics, and the mathematical sciences, a Parameter is a quantity that serves to relate functions and variables using a common variable (often t) when such a relationship would be difficult to explicate with an equation. In different contexts the term may have special uses.

· In a section on frequently misused words in his book The Writer's Art, James J. Kilpatrick quoted a letter from a correspondent, giving examples to illustrate the correct use of the word Parameter:

· A parametric equaliser is an audio filter that allows the frequency of maximum cut or boost to be set by one control, and the size of the cut or boost by another. These settings, the frequency level of the peak or trough, are two of the Parameters of a frequency response curve, and in a two-control equaliser they completely describe the curve.

Primary

A primary is the main physical body of a gravitationally-bound, multi-object system. This body contributes most of the mass of that system and will generally be located near its center of mass.

In the solar system, the Sun is the primary for all objects that orbit around it.

Temperature

Temperature is a physical property of matter that quantitatively expresses the common notions of hot and cold. Objects of low temperature are cold, while various degrees of higher temperatures are referred to as warm or hot. Quantitatively, temperature is measured with thermometers, which may be calibrated to a variety of temperature scales.

Lead

Lead refers to which set of legs, left or right, leads or advances forward to a greater extent when a quadruped animal is cantering, galloping, or leaping. The feet on the leading side touch the ground forward of its partner. On the "left lead", the animal's left legs lead.

Integral

Integration is an important concept in mathematics and, together with differentiation, is one of the two main operations in calculus. Given a function f of a real variable x and an interval [a, b] of the real line, the definite integral

is defined informally to be the net signed area of the region in the xy-plane bounded by the graph of f, the x-axis, and the vertical lines x = a and x = b.

The principles of integration were formulated independently by Isaac Newton and Gottfried Leibniz in the late 17th century.

Sensitivity

Sensitivity of explosives is the degree to which an explosive can be initiated by impact, heat, or friction.

Sensitivity, along with stability and brisance are three of the most significant properties of explosives that affect their use and application. All explosive compounds have a certain amount of energy required to initiate.

Uncertainty	Uncertainty is a term used in subtly different ways in a number of fields, including philosophy, physics, statistics, economics, finance, insurance, psychology, sociology, engineering, and information science. It applies to predictions of future events, to physical measurements already made, or to the unknown. In his seminal work Risk, Uncertainty, and Profit University of Chicago economist Frank Knight (1921) established the important distinction between risk and Uncertainty: Although the terms are used in various ways among the general public, many specialists in decision theory, statistics and other quantitative fields have defined Uncertainty and risk more specifically. Doug Hubbard defines Uncertainty and risk as: · Uncertainty: The lack of certainty, A state of having limited knowledge where it is impossible to exactly describe existing state or future outcome, more than one possible outcome. · Measurement of Uncertainty: A set of possible states or outcomes where probabilities are assigned to each possible state or outcome - this also includes the application of a probability density function to continuous variables · Risk: A state of Uncertainty where some possible outcomes have an undesired effect or significant loss. · Measurement of Risk: A set of measured uncertainties where some possible outcomes are losses, and the magnitudes of those losses - this also includes loss functions over continuous variables.
Reference	Hydrogen silsesquioxane is a negative tone electron-beam resist. Thin HSQ layers have been used to show very high resolution (~10 nm) feature sizes. Reference · http://cat.inist.fr/?aModele=afficheN'cpsidt=18807320 :'.

Specification	In regression analysis and related fields such as econometrics, specification is the process of converting a theory into a regression model. This process consists of selecting an appropriate functional form for the model and choosing which variables to include. Model specification is one of the first steps in regression analysis.
Spectrum	A spectrum is a condition that is not limited to a specific set of values but can vary infinitely within a continuum. The word saw its first scientific use within the field of optics to describe the rainbow of colors in visible light when separated using a prism; it has since been applied by analogy to many fields other than optics. Thus, one might talk about the spectrum of political opinion, or the spectrum of activity of a drug, or the autism spectrum.
Constraint	Curve fitting is the process of constructing a curve, that has the best fit to a series of data points, possibly subject to constraints. Curve fitting can involve either interpolation, where an exact fit to the data is required, or smoothing, in which a 'smooth' function is constructed that approximately fits the data. A related topic is regression analysis, which focuses more on questions of statistical inference such as how much uncertainty is present in a curve that is fit to data observed with random errors.
Circle	A circle is a simple shape of Euclidean geometry consisting of those points in a plane which are equidistant from a given point called the centre or center . The common distance of the points of a circle from its centre is called its radius. Circles are simple closed curves which divide the plane into two regions, an interior and an exterior.
Amplifier	Generally, an Amplifier or simply amp, is any device that changes, usually increases, the amplitude of a signal. The relationship of the input to the output of an Amplifier--usually expressed as a function of the input frequency--is called the transfer function of the Amplifier, and the magnitude of the transfer function is termed the gain. In popular use, the term usually describes an electronic Amplifier, in which the input 'signal' is usually a voltage or a current.

Rise time	In electronics, when describing a voltage or current step function, Rise time refers to the time required for a signal to change from a specified low value to a specified high value. Typically, these values are 10% and 90% of the step height. The output signal of a system is characterized also by fall time: both parameters depend on rise and fall times of input signal and on the characteristics of the system.

Lyapunov stability

In mathematics, the notion of Lyapunov stability occurs in the study of dynamical systems. In simple terms, if all solutions of the dynamical system that start out near an equilibrium point x_e stay near x_e forever, then x_e is Lyapunov stable. More strongly, if x_e is Lyapunov stable and all solutions that start out near x_e converge to x_e, then x_e is asymptotically stable.

State

In functional analysis, a state on a C*-algebra is a positive linear functional of norm 1. The set of states of a C*-algebra A, sometimes denoted by S(A), is always a convex set. The extremal points of S(A) are called pure states. If A has a multiplicative identity, S(A) is compact in the weak*-topology.

State space

In the theory of discrete dynamical systems, a state space is a directed graph where each possible state of a dynamical system is represented by a vertex, and there is a directed edge from a to b if and only if $f(a) = b$ where the function f defines the dynamical system.

State spaces are useful in computer science as a simple model of machines. Formally, it can be defined as a tuple [N, A, S, G] where:

- N is a set of states
- A is a set of arcs connecting the states
- S is a nonempty subset of N that contains start states
- G is a nonempty subset of N that contains the goal states.

The state space is what state space search searches in.

Perspective

Perspective, in context of vision and visual perception, is the way in which objects appear to the eye based on their spatial attributes; or their dimensions and the position of the eye relative to the objects. There are two main meanings of the term: linear perspective and aerial perspective.

Linear perspective

As objects become more distant they appear smaller because their visual angle decreases.

Stability	In probability theory, the stability of a random variable is the property that a linear combination of two independent copies of the variable has the same distribution, up to location and scale parameters. The distributions of random variables having this property are said to be "stable distributions". Results available in probability theory show that all possible distributions having this property are members of a four-parameter family of distributions.
Description	Description is one of four rhetorical modes (also known as modes of discourse), along with exposition, argumentation, and narration. Each of the rhetorical modes is present in a variety of forms and each has its own purpose and conventions. description is also the fiction-writing mode for transmitting a mental image of the particulars of a story.
Differential	A differential is a device, usually but not necessarily employing gears, capable of transmitting torque and rotation through three shafts, almost always used in one of two ways: in one way, it receives one input and provides two outputs--this is found in most automobiles--and in the other way, it combines two inputs to create an output that is the sum, difference, or average, of the inputs. In automobiles and other wheeled vehicles, the differential allows each of the driving roadwheels to rotate at different speeds, while for most vehicles supplying equal torque to each of them. Purpose A vehicle's wheels rotate at different speeds, mainly when turning corners.
Solution	In chemistry, a solution is a homogeneous mixture composed of two or more substances. In such a mixture, a solute is dissolved in another substance, known as a solvent. The solvent does the dissolving.
Standard	A technical standard is an established norm or requirement. It is usually a formal document that establishes uniform engineering or technical criteria, methods, processes and practices. In contrast, a custom, convention, company product, corporate standard, etc which becomes generally accepted and dominant is often called a de facto standard.

Transmission	A transmission or gearbox provides speed and torque conversions from a rotating power source to another device using gear ratios. In British English the term transmission refers to the whole drive train, including gearbox, clutch, prop shaft (for rear-wheel drive), differential and final drive shafts. In American English, however, the distinction is made that a gearbox is any device which converts speed and torque, whereas a transmission is a type of gearbox that can be "shifted" to dynamically change the speed:torque ratio, such as in a vehicle.
Dynamics	In the field of physics, the study of the causes of motion and changes in motion is dynamics. In other words the study of forces and why objects are in motion. Dynamics includes the study of the effect of torques on motion.
Step response	The step response of a system in a given initial state consists of the time evolution of its outputs when its control inputs are Heaviside step functions. In electronic engineering and control theory, step response is the time behaviour of the outputs of a general system when its inputs change from zero to one in a very short time. The concept can be extended to the abstract mathematical notion of a dynamical system using an evolution parameter.
DC motor	A DC motor is an electric motor that runs on direct current (DC) electricity. Brushed The brushed DC electric motor generates torque directly from DC power supplied to the motor by using internal commutation, stationary permanent magnets, and rotating electrical magnets. Like all electric motors or generators, torque is produced by the principle of Lorentz force, which states that any current-carrying conductor placed within an external magnetic field experiences a torque or force known as Lorentz force.
Block	A block in the context of extraction of petroleum, (and natural gas) signifies a sub-division of an underground reservoir for the purpose of administering a certain production allowable.
Block diagram	Block diagram is a diagram of a system, in which the principal parts or functions are represented by blocks connected by lines, that show the relationships of the blocks. They are heavily used in the engineering world in hardware design, software design, and process flow diagrams.

The block diagram is typically used for a higher level, less detailed description aimed more at understanding the overall concepts and less at understanding the details of implementation.

Implementation

Implementation is the realization of an application, or execution of a plan, idea, model, design, specification, standard, algorithm, or policy.

Computer Science

In computer science, an implementation is a realization of a technical specification or algorithm as a program, software component, or other computer system through programming and deployment. Many implementations may exist for a given specification or standard.

Transfer function

A transfer function is a mathematical representation, in terms of spatial or temporal frequency, of the relation between the input and output of a linear time-invariant system. With optical imaging devices, for example, it is the Fourier transform of the point spread function (hence a function of spatial frequency) i.e. the intensity distribution caused by a point object in the field of view.

Explanation

Transfer functions are commonly used in the analysis of systems such as single-input single-output filters, typically within the fields of signal processing, communication theory, and control theory.

LAPACK

LAPACK is a software library for numerical linear algebra. It provides routines for solving systems of linear equations and linear least squares, eigenvalue problems, and singular value decomposition. It also includes routines to implement the associated matrix factorizations such as LU, QR, Cholesky and Schur decomposition.

QR algorithm	In numerical linear algebra, the QR algorithm is an eigenvalue algorithm: that is, a procedure to calculate the eigenvalues and eigenvectors of a matrix. The QR transformation was developed in the late 1950s by John G.F. Francis (England) and by Vera N. Kublanovskaya (USSR), working independently. The basic idea is to perform a QR decomposition, writing the matrix as a product of an orthogonal matrix and an upper triangular matrix, multiply the factors in the other order, and iterate.
Estimator	In statistics, an estimator is a rule for calculating an estimate of a given quantity based on observed data: thus the rule and its result (the estimate) are distinguished This is in contrast to an interval estimator, where the result would be a range of plausible values (or vectors or functions).
Laplace transform	In mathematics, the Laplace transform is a widely used integral transform. Denoted $\mathcal{L}\{f(t)\}$, it is a linear operator of a function f(t) with a real argument t (t ≥ 0) that transforms it to a function F(s) with a complex argument s. This transformation is essentially bijective for the majority of practical uses; the respective pairs of f(t) and F(s) are matched in tables.
Forcing function	In a system of differential equations used to describe a time-dependent process, a forcing function is a function that appears in the equations and is only a function of time, not of any of the other variables. In effect, it is a constant for each value of t.
Frequency	Frequency is the number of occurrences of a repeating event per unit time. It is also referred to as temporal frequency. The period is the duration of one cycle in a repeating event, so the period is the reciprocal of the frequency.
Uncertainty	Uncertainty is a term used in subtly different ways in a number of fields, including philosophy, physics, statistics, economics, finance, insurance, psychology, sociology, engineering, and information science. It applies to predictions of future events, to physical measurements already made, or to the unknown. In his seminal work Risk, Uncertainty, and Profit University of Chicago economist Frank Knight (1921) established the important distinction between risk and Uncertainty: Although the terms are used in various ways among the general public, many specialists in decision theory, statistics and other quantitative fields have defined Uncertainty and risk more specifically. Doug Hubbard defines Uncertainty and risk as: · Uncertainty: The lack of certainty, A state of having limited knowledge where it is impossible to exactly describe existing state or future outcome, more than one possible outcome.

· Measurement of Uncertainty: A set of possible states or outcomes where probabilities are assigned to each possible state or outcome - this also includes the application of a probability density function to continuous variables

· Risk: A state of Uncertainty where some possible outcomes have an undesired effect or significant loss.

· Measurement of Risk: A set of measured uncertainties where some possible outcomes are losses, and the magnitudes of those losses - this also includes loss functions over continuous variables.

Compensation

In engineering, compensation is planning for side effects or other unintended issues in a design. The design of an invention can itself also be to compensate for some other existing issue or exception.

One example is in a voltage-controlled crystal oscillator (VCXO), which is normally affected not only by voltage, but to a lesser extent by temperature.

Pendulum

Derivations from Pendulum.

Analysis of a simple gravity pendulum

To begin, we shall make three assumptions about the simple pendulum:

- The rod/string/cable on which the bob is swinging is massless, does not stretch, and always remains taut.
- The bob is a point mass.
- Motion occurs in a 2-dimensional plane, i.e. pendulum does not swing into and out of the page.

Consider Figure 1, showing the forces acting on a simple pendulum. Note that the path of the pendulum sweeps out an arc of a circle.

Oscillator	An oscillator is a technical analysis indicator that varies over time within a band (above and below a center line, or between set levels). Oscillators are used to discover short-term overbought or oversold conditions. Common oscillators are MACD, ROC, RSI, CCI.
Placement	Placement is an essential step in electronic design automation - the portion of the physical design flow that assigns exact locations for various circuit components within the chip's core area. An inferior placement assignment will not only affect the chip's performance but might also make it nonmanufacturable by producing excessive wirelength, which is beyond available routing resources. Consequently, a placer must perform the assignment while optimizing a number of objectives to ensure that a circuit meets its performance demands.
Reference	Hydrogen silsesquioxane is a negative tone electron-beam resist. Thin HSQ layers have been used to show very high resolution (~10 nm) feature sizes. Reference · http://cat.inist.fr/?aModele=afficheN'cpsidt=18807320 :'.
Pole	In the mathematical field of complex analysis, a pole of a meromorphic function is a certain type of singularity that behaves like the singularity of $\frac{1}{z^n}$ at z = 0. This means that, in particular, a pole of the function f(z) is a point a such that f(z) approaches infinity as z approaches a.

Formally, suppose U is an open subset of the complex plane C, a is an element of U and f : U {a} → C is a function which is holomorphic over its domain. If there exists a holomorphic function g : U → C and a positive integer n, such that for all z in U {a}

$$f(z) = \frac{g(z)}{(z-a)^n}$$

holds, then a is called a pole of f.

Tension

In physics, tension is the magnitude of the pulling force exerted by a string, cable, chain, or similar object on another object. It is the opposite of compression. As tension is the magnitude of a force, it is measured in newtons (or sometimes pounds-force) and is always measured parallel to the string on which it applies.

Step response

The step response of a system in a given initial state consists of the time evolution of its outputs when its control inputs are Heaviside step functions. In electronic engineering and control theory, step response is the time behaviour of the outputs of a general system when its inputs change from zero to one in a very short time. The concept can be extended to the abstract mathematical notion of a dynamical system using an evolution parameter.

Rise time

In electronics, when describing a voltage or current step function, Rise time refers to the time required for a signal to change from a specified low value to a specified high value. Typically, these values are 10% and 90% of the step height. The output signal of a system is characterized also by fall time: both parameters depend on rise and fall times of input signal and on the characteristics of the system.

Simulink

Simulink, developed by MathWorks, is a commercial tool for modeling, simulating and analyzing multidomain dynamic systems. Its primary interface is a graphical block diagramming tool and a customizable set of block libraries. It offers tight integration with the rest of the MATLAB environment and can either drive MATLAB or be scripted from it.

Speed

In kinematics, the speed of an object is the magnitude of its velocity (the rate of change of its position); it is thus a scalar quantity. The average speed of an object in an interval of time is the distance traveled by the object divided by the duration of the interval; the instantaneous speed is the limit of the average speed as the duration of the time interval approaches zero .

Like velocity, speed has the dimensions of a length divided by a time; the SI unit of speed is the meter per second, but the most usual unit of speed in everyday usage is the kilometer per hour or, in the USA and the UK, miles per hour.

Inverted pendulum

An inverted pendulum is a pendulum which has its mass above its pivot point. It is often implemented with the pivot point mounted on a cart that can move horizontally and may be called a cart and pole. Whereas a normal pendulum is stable when hanging downwards, an inverted pendulum is inherently unstable, and must be actively balanced in order to remain upright, either by applying a torque at the pivot point or by moving the pivot point horizontally as part of a feedback system.

Nyquist plot

A Nyquist plot is used in automatic control and signal processing for assessing the stability of a system with feedback. It is represented by a graph in polar coordinates in which the gain and phase of a frequency response are plotted. The plot of these phasor quantities shows the phase as the angle and the magnitude as the distance from the origin.

Tape drive

A tape drive is a data storage device that reads and writes data on a magnetic tape. It is typically used for offline, archival data storage. Tape media generally has a favorable unit cost and long archival stability.

Regulator

In automatic control, a regulator is a device which has the function of maintaining a designated characteristic. It performs the activity of managing or maintaining a range of values in a machine. The measurable property of a device is managed closely by specified conditions or an advance set value; or it can be a variable according to a predetermined arrangement scheme.

Regulator

In automatic control, a regulator is a device which has the function of maintaining a designated characteristic. It performs the activity of managing or maintaining a range of values in a machine. The measurable property of a device is managed closely by specified conditions or an advance set value; or it can be a variable according to a predetermined arrangement scheme.

Phase margin	In electronic amplifiers, phase margin is the difference between the phase, measured in degrees, of an amplifier's output signal (relative to its input) and 180°, as a function of frequency. The Phase margin is taken as positive at frequencies below where the open-loop phase first crosses 180°, i.e. the signal becomes inverted, or antiphase; it is negative beyond. In negative feedback, a negative Phase margin at a frequency where the loop gain exceeds unity guarantees instability, thus positive Phase margin is a "safety margin" that ensures proper operation of an amplifier and, more generally, active filters, under various loads.
Inductor	An inductor is a passive electrical component that can store energy in a magnetic field created by the electric current passing through it. An inductor's ability to store magnetic energy is measured by its inductance, in units of henries. Typically an inductor is a conducting wire shaped as a coil; the loops help to create a strong magnetic field inside the coil due to Ampere's Law.
Observer	In quantum mechanics, "observation" is synonymous with quantum measurement and "observer" with a measurement apparatus and observable with what can be measured. Thus the quantum mechanical observer does not necessarily present or solve any problems over and above the (admittedly difficult) issue of measurement in quantum mechanics. The quantum mechanical observer is also intimately tied to the issue of observer effect.
Controller	In control theory, a controller is a device which monitors and affects the operational conditions of a given dynamical system. The operational conditions are typically referred to as output variables of the system which can be affected by adjusting certain input variables. For example, the heating system of a house can be equipped with a thermostat (controller) for sensing air temperature (output variable) which can turn on or off a furnace or heater when the air temperature becomes too low or too high.
Mechanism	A mechanism is a combination of resistant bodies, so interconnected that by applying force or motion to one or more of those bodies, some of those bodies are caused to perform desired work accompanied by desired motions. Mechanisms are used to convert one type of motion into another. The connections present in a mechanism are called kinematic pairs.
Bode plot	A Bode plot is a graph of the transfer function of a linear, time-invariant system versus frequency, plotted with a log-frequency axis, to show the system's frequency response. It is usually a combination of a Bode magnitude plot, expressing the magnitude of the frequency response gain, and a Bode phase plot, expressing the frequency response phase shift. Overview

Among his several important contributions to circuit theory and control theory, engineer Hendrik Wade Bode (1905-1982), while working at Bell Labs in the United States in the 1930s, devised a simple but accurate method for graphing gain and phase-shift plots.

Design methods

Design Methods is a broad area that focuses on:

- Divergence - Exploring possibilities and constraints of inherited situations by applying critical thinking through qualitative and quantitative research methods to create new understanding (problem space) toward better design solutions
- Transformation - Redefining specifications of design solutions which can lead to better guidelines for traditional and contemporary design activities (architecture, graphic, industrial, information, interaction, et al). and/or multidisciplinary response
- Convergence - Prototyping possible scenarios for better design solutions that incrementally or significantly improve the originally inherited situation
- Sustainability - Managing the process of exploring, redefining and prototyping of design solutions continually over time
- Articulation - the visual relationship between the parts and the whole.

The goal of design methods is to gain key insights or unique essential truths resulting in more holistic solutions in order to achieve better experiences for users with products, services, environments and systems they rely upon. Insight, in this case, is clear and deep investigation of a situation through design methods, thereby grasping the inner nature of things intuitively.

Frequency response

Frequency response is the measure of any system's output spectrum in response to an input signal. In the audible range it is usually referred to in connection with electronic amplifiers, microphones and loudspeakers. Radio spectrum frequency response can refer to measurements of coaxial cables, category cables, video switchers and wireless communications devices.

Continuous

In probability theory, a probability distribution is called continuous if its cumulative distribution function is continuous. This is equivalent to saying that for random variables X with the distribution in question, Pr[X = a] = 0 for all real numbers a, i.e.: the probability that X attains the value a is zero, for any number a. If the distribution of X is continuous then X is called a continuous random variable.

Discrete system

A discrete system is a system with a countable number of states. Discrete systems may be contrasted with continuous systems, which may also be called analog systems. A final discrete system is often modeled with a directed graph and is analyzed for correctness and complexity according to computational theory.

Integral

Integration is an important concept in mathematics and, together with differentiation, is one of the two main operations in calculus. Given a function f of a real variable x and an interval [a, b] of the real line, the definite integral

is defined informally to be the net signed area of the region in the xy-plane bounded by the graph of f, the x-axis, and the vertical lines x = a and x = b.

The principles of integration were formulated independently by Isaac Newton and Gottfried Leibniz in the late 17th century.

Tracking

In particle physics, the tracking is the act of measuring the direction and magnitude of charged particles momenta. The particles entering a tracker (the device used for tracking), release part of their energy in the device: the tracker has to be finely segmented in order to be able to reconstruct with good precision where the particle passed. Since the tracking is usually made in a region where a magnetic field is present, it is possible to reconstruct part of the helix made by the particle inside the tracker (that is called track), and from the track parameters, and by knowing the mass of the particle under study (which is known by the use of particle identification), it is possible to reconstruct the actual direction and magnitude of the particle momenta.

Signal

In the fields of communications, signal processing, and in electrical engineering more generally, a signal is any time-varying or spatial-varying quantity.

In the physical world, any quantity measurable through time or over space can be taken as a signal. Within a complex society, any set of human information or machine data can also be taken as a signal.

Signal

In the fields of communications, signal processing, and in electrical engineering more generally, a signal is any time-varying or spatial-varying quantity.

In the physical world, any quantity measurable through time or over space can be taken as a signal. Within a complex society, any set of human information or machine data can also be taken as a signal.

Steady state

In chemistry, a steady state is a situation in which all state variables are constant in spite of ongoing processes that strive to change them. For an entire system to be at steady state, i.e. for all state variables of a system to be constant, there must be a flow through the system (compare mass balance). One of the most simple examples of such a system is the case of a bathtub with the tap open but without the bottom plug: after a certain time the water flows in and out at the same rate, so the water level (the state variable being Volume) stabilizes and the system is at steady state.

Recovery

Recovery is a process by which deformed grains can reduce their stored energy by the removal or rearrangement of defects in their crystal structure. These defects, primarily dislocations, are introduced by plastic deformation of the material and act to increase the yield strength of a material. Since recovery reduces the dislocation density the process is normally accompanied by a reduction in a materials strength and a simultaneous increase in the ductility.

Semiconductor

A semiconductor is a material with electrical conductivity due to electron flow (as opposed to ionic conductivity) intermediate in magnitude between that of a conductor and an insulator. This means a conductivity roughly in the range of 10^3 to 10^{-8} siemens per centimeter. Semiconductor materials are the foundation of modern electronics, including radio, computers, telephones, and many other devices.

Manufacturing

Manufacturing is the use of machines, tools and labor to produce goods for use or sale. The term may refer to a range of human activity, from handicraft to high tech, but is most commonly applied to industrial production, in which raw materials are transformed into finished goods on a large scale. Such finished goods may be used for manufacturing other, more complex products, such as aircraft, household appliances or automobiles, or sold to wholesalers, who in turn sell them to retailers, who then sell them to end users - the "consumers".

Series

A series is the sum of the terms of a sequence. Finite sequences and series have defined first and last terms, whereas infinite sequences and series continue indefinitely.

In mathematics, given an infinite sequence of numbers { a_n }, a series is informally the result of adding all those terms together: $a_1 + a_2 + a_3 + \cdots$.

Transfer function

A transfer function is a mathematical representation, in terms of spatial or temporal frequency, of the relation between the input and output of a linear time-invariant system. With optical imaging devices, for example, it is the Fourier transform of the point spread function (hence a function of spatial frequency) i.e. the intensity distribution caused by a point object in the field of view.

Explanation

Transfer functions are commonly used in the analysis of systems such as single-input single-output filters, typically within the fields of signal processing, communication theory, and control theory.

Polynomial

The polynomial hyperelastic material model is a phenomenological model of rubber elasticity. In this model, the strain energy density function is of the form of a polynomial in the two invariants I_1, I_2 of the left Cauchy-Green deformation tensor.

The strain energy density function for the polynomial model is

$$W = \sum_{i,j=0}^{n} C_{ij}(I_1 - 3)^i (I_2 - 3)^j$$

where C_{ij} are material constants and $C_{00} = 0$.

Polynomial

The polynomial hyperelastic material model is a phenomenological model of rubber elasticity. In this model, the strain energy density function is of the form of a polynomial in the two invariants I_1, I_2 of the left Cauchy-Green deformation tensor.

The strain energy density function for the polynomial model is

$$W = \sum_{i,j=0}^{n} C_{ij}(I_1 - 3)^i (I_2 - 3)^j$$

where C_{ij} are material constants and $C_{00} = 0$.

Digital

A digital system is a data technology that uses discrete (discontinuous) values. By contrast, non-digital systems use a continuous range of values to represent information. Although digital representations are discrete, the information represented can be either discrete, such as numbers, letters or icons, or continuous, such as sounds, images, and other measurements of continuous systems.

Heat exchanger

A heat exchanger is a device built for efficient heat transfer from one medium to another. The media may be separated by a solid wall, so that they never mix, or they may be in direct contact. They are widely used in space heating, refrigeration, air conditioning, power plants, chemical plants, petrochemical plants, petroleum refineries, natural gas processing, and sewage treatment.

Discrete signal	A discrete signal is a time series consisting of a sequence of quantities. In other words, it is a time series that is a function over a domain of discrete integers. Each value in the sequence is called a sample.

Pulse

In signal processing, the term pulse has the following meanings:

1. A rapid, transient change in the amplitude of a signal from a baseline value to a higher or lower value, followed by a rapid return to the baseline value.
2. A rapid change in some characteristic of a signal, e.g., phase or frequency, from a baseline value to a higher or lower value, followed by a rapid return to the baseline value.

Pulse shapes

Pulse shapes can arise out of a process called pulse-shaping. Optimum pulse shape depends on the application.

Rectangular pulse

These can be found in pulse waves, square waves, boxcar functions, and rectangular functions.

Digital

A digital system is a data technology that uses discrete (discontinuous) values. By contrast, non-digital systems use a continuous range of values to represent information. Although digital representations are discrete, the information represented can be either discrete, such as numbers, letters or icons, or continuous, such as sounds, images, and other measurements of continuous systems.

Period

In the periodic table of the elements, elements are arranged in a series of rows (or periods) so that those with similar properties appear in vertical columns. Elements of the same period have the same number of electron shells; with each group across a period, the elements have one more proton and electron and become less metallic. This arrangement reflects the periodic recurrence of similar properties as the atomic number increases.

Running	Running is a means of terrestrial locomotion allowing a human or an animal to move rapidly on foot. It is simply defined in athletics terms as a gait in which at regular points during the running cycle both feet are off the ground. This is in contrast to walking, where one foot is always in contact with the ground, the legs are kept mostly straight and the center of gravity vaults over the legs in an inverted pendulum fashion.
Signal	In the fields of communications, signal processing, and in electrical engineering more generally, a signal is any time-varying or spatial-varying quantity. In the physical world, any quantity measurable through time or over space can be taken as a signal. Within a complex society, any set of human information or machine data can also be taken as a signal.
Continuous	In probability theory, a probability distribution is called continuous if its cumulative distribution function is continuous. This is equivalent to saying that for random variables X with the distribution in question, Pr[X = a] = 0 for all real numbers a, i.e.: the probability that X attains the value a is zero, for any number a. If the distribution of X is continuous then X is called a continuous random variable.
Block	A block in the context of extraction of petroleum, (and natural gas) signifies a sub-division of an underground reservoir for the purpose of administering a certain production allowable.
Block diagram	Block diagram is a diagram of a system, in which the principal parts or functions are represented by blocks connected by lines, that show the relationships of the blocks. They are heavily used in the engineering world in hardware design, software design, and process flow diagrams. The block diagram is typically used for a higher level, less detailed description aimed more at understanding the overall concepts and less at understanding the details of implementation.
Discrete system	A discrete system is a system with a countable number of states. Discrete systems may be contrasted with continuous systems, which may also be called analog systems. A final discrete system is often modeled with a directed graph and is analyzed for correctness and complexity according to computational theory.

Chapter 8. Digital Control

Dynamics	In the field of physics, the study of the causes of motion and changes in motion is dynamics. In other words the study of forces and why objects are in motion. Dynamics includes the study of the effect of torques on motion.
Laplace transform	In mathematics, the Laplace transform is a widely used integral transform. Denoted $\mathcal{L}\{f(t)\}$, it is a linear operator of a function f(t) with a real argument t (t ≥ 0) that transforms it to a function F(s) with a complex argument s. This transformation is essentially bijective for the majority of practical uses; the respective pairs of f(t) and F(s) are matched in tables.
Simulink	Simulink, developed by MathWorks, is a commercial tool for modeling, simulating and analyzing multidomain dynamic systems. Its primary interface is a graphical block diagramming tool and a customizable set of block libraries. It offers tight integration with the rest of the MATLAB environment and can either drive MATLAB or be scripted from it.
Discrete time	Discrete time is non-continuous time. Sampling at non-continuous times results in discrete-time samples. For example, a newspaper may report the price of crude oil once every 24 hours.
Inversion	In meteorology, an inversion is a deviation from the normal change of an atmospheric property with altitude. It almost always refers to a temperature inversion, i.e., an increase in temperature with height, or to the layer (inversion layer) within which such an increase occurs. An inversion can lead to pollution such as smog being trapped close to the ground, with possible adverse effects on health.
Continuous signal	A continuous signal is a varying quantity (a signal) whose domain, which is often time, is a continuum (e.g., a connected interval of the reals). That is, the function's domain is an uncountable set. The function itself need not be continuous.
Aliasing	In signal processing and related disciplines, Aliasing refers to an effect that causes different signals to become indistinguishable (or aliases of one another) when sampled. It also refers to the distortion or artifact that results when the signal reconstructed from samples is different than the original continuous signal.

When a digital image is viewed, a reconstruction - also known as an interpolation - is performed by a display or printer device, and by the eyes and the brain.

Damping

In physics, damping is any effect that tends to reduce the amplitude of oscillations in an oscillatory system, particularly the harmonic oscillator.

In mechanics, friction is one such damping effect. For many purposes the frictional force F_f can be modeled as being proportional to the velocity v of the object:

$$F_{\mathrm{f}} = -cv \,,$$

where c is the viscous damping coefficient, given in units of Newton-seconds per meter.

Final value theorem

In mathematical analysis, the final value theorem is one of several similar theorems used to relate frequency domain expressions to the time domain behavior as time approaches infinity. A final value theorem allows the time domain behavior to be directly calculated by taking a limit of a frequency domain expression, as opposed to converting to a time domain expression and taking its limit.

Mathematically, if

$$\lim_{t \to \infty} f(t)$$

exists, then

$$\lim_{t \to \infty} f(t) = \lim_{s \to 0} sF(s)$$

where F(s) is the Laplace transform of the signal f(t).

Folding

In chemistry folding is the process by which a molecule assumes its shape or conformation. The process can also be described as intramolecular self-assembly where the molecule is directed to form a specific shape through noncovalent interactions, such as hydrogen bonding, metal coordination, hydrophobic forces, van der Waals forces, pi-pi interactions, and/or electrostatic effects.

The most active area of interest in the folding of molecules is the process of protein folding, which is the shape that is assumed by a specific sequence of amino acids in a protein.

Frequency

Frequency is the number of occurrences of a repeating event per unit time. It is also referred to as temporal frequency. The period is the duration of one cycle in a repeating event, so the period is the reciprocal of the frequency.

Sequence

In music, a sequence is the immediate restatement of a motif or longer melodic (or harmonic) passage at a higher or lower pitch in the same voice. It is one of the most common and simple methods of elaborating a melody in eighteenth and nineteenth century classical music (Classical period (music) and Romantic music). Characteristics of sequences:

- Two segments, usually no more than three or four
- Usually only one direction: continuingly higher or lower
- Segments continue by same interval distance

It is possible for melody or harmony to form a sequence without the other participating.

Implementation

Implementation is the realization of an application, or execution of a plan, idea, model, design, specification, standard, algorithm, or policy.

Computer Science

In computer science, an implementation is a realization of a technical specification or algorithm as a program, software component, or other computer system through programming and deployment. Many implementations may exist for a given specification or standard.

Approximation

An approximation is an inexact representation of something that is still close enough to be useful. Although approximation is most often applied to numbers, it is also frequently applied to such things as mathematical functions, shapes, and physical laws.

Approximations may be used because incomplete information prevents use of exact representations.

Controller

In control theory, a controller is a device which monitors and affects the operational conditions of a given dynamical system. The operational conditions are typically referred to as output variables of the system which can be affected by adjusting certain input variables. For example, the heating system of a house can be equipped with a thermostat (controller) for sensing air temperature (output variable) which can turn on or off a furnace or heater when the air temperature becomes too low or too high.

Sampling

In signal processing, sampling is the reduction of a continuous signal to a discrete signal. A common example is the conversion of a sound wave (a continuous-time signal) to a sequence of samples (a discrete-time signal).

A sample refers to a value or set of values at a point in time and/or space.

EPROM

An EPROM or erasable programmable read only memory, is a type of memory chip that retains its data when its power supply is switched off. In other words, it is non-volatile. It is an array of floating-gate transistors individually programmed by an electronic device that supplies higher voltages than those normally used in digital circuits.

Fixed point

In mathematics, a fixed point of a function is a point that is mapped to itself by the function. A set of fixed points is sometimes called a fixed set. That is to say, x is a fixed point of the function f if and only if f(x) = x.

Arithmetic

Arithmetic or Arithmetics is the oldest and most elementary branch of mathematics, used by almost everyone, for tasks ranging from simple day-to-day counting to advanced science and business calculations. It involves the study of quantity, especially as the result of combining numbers. In common usage, it refers to the simpler properties when using the traditional operations of addition, subtraction, multiplication and division with smaller values of numbers.

Effectiveness

Effectiveness means the capability of producing an effect.

In mathematics, effective is sometimes used as a synonym of algorithmically computable.

In physics, an effective theory is, similar to a phenomenological theory, a framework intended to explain certain (observed) effects without the claim that the theory correctly models the underlying (unobserved) processes.

Tracking

In particle physics, the tracking is the act of measuring the direction and magnitude of charged particles momenta. The particles entering a tracker (the device used for tracking), release part of their energy in the device: the tracker has to be finely segmented in order to be able to reconstruct with good precision where the particle passed. Since the tracking is usually made in a region where a magnetic field is present, it is possible to reconstruct part of the helix made by the particle inside the tracker (that is called track), and from the track parameters, and by knowing the mass of the particle under study (which is known by the use of particle identification), it is possible to reconstruct the actual direction and magnitude of the particle momenta.

Control system

A control system is a device or set of devices to manage, command, direct or regulate the behavior of other devices or systems.

There are two common classes of control systems, with many variations and combinations: logic or sequential controls, and feedback or linear controls. There is also fuzzy logic, which attempts to combine some of the design simplicity of logic with the utility of linear control.

Transfer function

A transfer function is a mathematical representation, in terms of spatial or temporal frequency, of the relation between the input and output of a linear time-invariant system. With optical imaging devices, for example, it is the Fourier transform of the point spread function (hence a function of spatial frequency) i.e. the intensity distribution caused by a point object in the field of view.

Explanation

Transfer functions are commonly used in the analysis of systems such as single-input single-output filters, typically within the fields of signal processing, communication theory, and control theory.

Integral

Integration is an important concept in mathematics and, together with differentiation, is one of the two main operations in calculus. Given a function f of a real variable x and an interval [a, b] of the real line, the definite integral

is defined informally to be the net signed area of the region in the xy-plane bounded by the graph of f, the x-axis, and the vertical lines x = a and x = b.

The principles of integration were formulated independently by Isaac Newton and Gottfried Leibniz in the late 17th century.

Compensation

In engineering, compensation is planning for side effects or other unintended issues in a design. The design of an invention can itself also be to compensate for some other existing issue or exception.

One example is in a voltage-controlled crystal oscillator (VCXO), which is normally affected not only by voltage, but to a lesser extent by temperature.

Step response

The step response of a system in a given initial state consists of the time evolution of its outputs when its control inputs are Heaviside step functions. In electronic engineering and control theory, step response is the time behaviour of the outputs of a general system when its inputs change from zero to one in a very short time. The concept can be extended to the abstract mathematical notion of a dynamical system using an evolution parameter.

Series

A series is the sum of the terms of a sequence. Finite sequences and series have defined first and last terms, whereas infinite sequences and series continue indefinitely.

In mathematics, given an infinite sequence of numbers $\{ a_n \}$, a series is informally the result of adding all those terms together: $a_1 + a_2 + a_3 + \cdots$.

Perspective

Perspective, in context of vision and visual perception, is the way in which objects appear to the eye based on their spatial attributes; or their dimensions and the position of the eye relative to the objects. There are two main meanings of the term: linear perspective and aerial perspective.

Linear perspective

As objects become more distant they appear smaller because their visual angle decreases.

Torque	Torque, is the tendency of a force to rotate an object about an axis, fulcrum, or pivot. Just as a force is a push or a pull, a torque can be thought of as a twist. Loosely speaking, torque is a measure of the turning force on an object such as a bolt or a flywheel.
Dynamics	In the field of physics, the study of the causes of motion and changes in motion is dynamics. In other words the study of forces and why objects are in motion. Dynamics includes the study of the effect of torques on motion.
Linearization	In mathematics and its applications, linearization refers to finding the linear approximation to a function at a given point. In the study of dynamical systems, linearization is a method for assessing the local stability of an equilibrium point of a system of nonlinear differential equations or discrete dynamical systems. This method is used in fields such as engineering, physics, economics, and ecology.
Differential	A differential is a device, usually but not necessarily employing gears, capable of transmitting torque and rotation through three shafts, almost always used in one of two ways: in one way, it receives one input and provides two outputs--this is found in most automobiles--and in the other way, it combines two inputs to create an output that is the sum, difference, or average, of the inputs. In automobiles and other wheeled vehicles, the differential allows each of the driving roadwheels to rotate at different speeds, while for most vehicles supplying equal torque to each of them. Purpose A vehicle's wheels rotate at different speeds, mainly when turning corners.
Signal	In the fields of communications, signal processing, and in electrical engineering more generally, a signal is any time-varying or spatial-varying quantity.

In the physical world, any quantity measurable through time or over space can be taken as a signal. Within a complex society, any set of human information or machine data can also be taken as a signal.

Solution

In chemistry, a solution is a homogeneous mixture composed of two or more substances. In such a mixture, a solute is dissolved in another substance, known as a solvent. The solvent does the dissolving.

Motion

In physics, motion is a change in position of an object with respect to time. Change in action is the result of an unapplied force. Motion is typically described in terms of velocity, acceleration, displacement, and time.

Nonlinear

In mathematics, a nonlinear system is a system which is not linear, that is, a system which does not satisfy the superposition principle, a nonlinear system is any problem where the variable(s) to be solved for cannot be written as a linear combination of independent components. A nonhomogeneous system, which is linear apart from the presence of a function of the independent variables, is nonlinear according to a strict definition, but such systems are usually studied alongside linear systems, because they can be transformed to a linear system of multiple variables.

Pendulum

Derivations from Pendulum.

Analysis of a simple gravity pendulum

To begin, we shall make three assumptions about the simple pendulum:

- The rod/string/cable on which the bob is swinging is massless, does not stretch, and always remains taut.
- The bob is a point mass.
- Motion occurs in a 2-dimensional plane, i.e. pendulum does not swing into and out of the page.

Consider Figure 1, showing the forces acting on a simple pendulum. Note that the path of the pendulum sweeps out an arc of a circle.

Chapter 9. Nonlinear Systems

Frequency	Frequency is the number of occurrences of a repeating event per unit time. It is also referred to as temporal frequency. The period is the duration of one cycle in a repeating event, so the period is the reciprocal of the frequency.
Frequency response	Frequency response is the measure of any system's output spectrum in response to an input signal. In the audible range it is usually referred to in connection with electronic amplifiers, microphones and loudspeakers. Radio spectrum frequency response can refer to measurements of coaxial cables, category cables, video switchers and wireless communications devices.
Relay	A relay is an electrically operated switch. Many relays use an electromagnet to operate a switching mechanism mechanically, but other operating principles are also used. Relays are used where it is necessary to control a circuit by a low-power signal (with complete electrical isolation between control and controlled circuits), or where several circuits must be controlled by one signal.
Saturation	In chemistry, saturation has six different meanings, all based on reaching a maximum capacity 1. In physical chemistry, saturation is the point at which a solution of a substance can dissolve no more of that substance and additional amounts of it will appear as a precipitate. This point of maximum concentration, the saturation point, depends on the temperature of the liquid as well as the chemical nature of the substances involved. This can be used in the process of recrystallisation to purify a chemical: it is dissolved to the point of saturation in hot solvent, then as the solvent cools and the solubility decreases, excess solute precipitates.
Stability	In probability theory, the stability of a random variable is the property that a linear combination of two independent copies of the variable has the same distribution, up to location and scale parameters. The distributions of random variables having this property are said to be "stable distributions". Results available in probability theory show that all possible distributions having this property are members of a four-parameter family of distributions.
Block	A block in the context of extraction of petroleum, (and natural gas) signifies a sub-division of an underground reservoir for the purpose of administering a certain production allowable.

Block diagram

Block diagram is a diagram of a system, in which the principal parts or functions are represented by blocks connected by lines, that show the relationships of the blocks. They are heavily used in the engineering world in hardware design, software design, and process flow diagrams.

The block diagram is typically used for a higher level, less detailed description aimed more at understanding the overall concepts and less at understanding the details of implementation.

Compensation

In engineering, compensation is planning for side effects or other unintended issues in a design. The design of an invention can itself also be to compensate for some other existing issue or exception.

One example is in a voltage-controlled crystal oscillator (VCXO), which is normally affected not only by voltage, but to a lesser extent by temperature.

Estimator

In statistics, an estimator is a rule for calculating an estimate of a given quantity based on observed data: thus the rule and its result (the estimate) are distinguished This is in contrast to an interval estimator, where the result would be a range of plausible values (or vectors or functions).

Controller

In control theory, a controller is a device which monitors and affects the operational conditions of a given dynamical system. The operational conditions are typically referred to as output variables of the system which can be affected by adjusting certain input variables. For example, the heating system of a house can be equipped with a thermostat (controller) for sensing air temperature (output variable) which can turn on or off a furnace or heater when the air temperature becomes too low or too high.

Integral

Integration is an important concept in mathematics and, together with differentiation, is one of the two main operations in calculus. Given a function f of a real variable x and an interval [a, b] of the real line, the definite integral

is defined informally to be the net signed area of the region in the xy-plane bounded by the graph of f, the x-axis, and the vertical lines x = a and x = b.

The principles of integration were formulated independently by Isaac Newton and Gottfried Leibniz in the late 17th century.

Schmitt trigger

In electronics, Schmitt trigger is a generic name of threshold circuits with positive feedback having a loop gain > 1. The circuit is named "trigger" because the output retains its value until the input changes sufficiently to trigger a change: in the non-inverting configuration, when the input is higher than a certain chosen threshold, the output is high; when the input is below a different (lower) chosen threshold, the output is low; when the input is between the two, the output retains its value. This dual threshold action is called hysteresis and implies that the Schmitt trigger possess memory and can act as a bistable circuit (latch). There is a close relation between the two kinds of circuits that actually are the same: a Schmitt trigger can be converted into a latch and v.v., a latch can be converted into a Schmitt trigger.

Lyapunov stability

In mathematics, the notion of Lyapunov stability occurs in the study of dynamical systems. In simple terms, if all solutions of the dynamical system that start out near an equilibrium point x_e stay near x_e forever, then x_e is Lyapunov stable. More strongly, if x_e is Lyapunov stable and all solutions that start out near x_e converge to x_e, then x_e is asymptotically stable.

Direct methods

In crystallography, direct methods are a family of methods for estimating the phases of the Fourier transform of the scattering density from the corresponding magnitudes. The methods generally exploit constraints or statistical correlations between the phases of different Fourier components that result from the fact that the scattering density must be a positive real number.

In two dimensions, it is relatively easy to solve the phase problem directly, but not so in three dimensions.

Nonlinear system

In mathematics, a nonlinear system is a system which is not linear, that is, a system which does not satisfy the superposition principle, or whose output is not directly proportional to its input. Less technically, a nonlinear system is any problem where the variable(s) to be solved for cannot be written as a linear combination of independent components. A nonhomogeneous system, which is linear apart from the presence of a function of the independent variables, is nonlinear according to a strict definition, but such systems are usually studied alongside linear systems, because they can be transformed to a linear system of multiple variables.

Circle	A circle is a simple shape of Euclidean geometry consisting of those points in a plane which are equidistant from a given point called the centre or center . The common distance of the points of a circle from its centre is called its radius. Circles are simple closed curves which divide the plane into two regions, an interior and an exterior.
Perspective	Perspective, in context of vision and visual perception, is the way in which objects appear to the eye based on their spatial attributes; or their dimensions and the position of the eye relative to the objects. There are two main meanings of the term: linear perspective and aerial perspective. Linear perspective As objects become more distant they appear smaller because their visual angle decreases.
Governor	A governor, is a device used to measure and regulate the speed of a machine, such as an engine. A classic example is the centrifugal governor, also known as the Watt or fly-ball governor, which uses weights mounted on spring-loaded arms to determine how fast a shaft is spinning, and then uses proportional control to regulate the shaft speed. History Centrifugal governors were used to regulate the distance and pressure between millstones in windmills since the 17th century.

Control system

A control system is a device or set of devices to manage, command, direct or regulate the behavior of other devices or systems.

There are two common classes of control systems, with many variations and combinations: logic or sequential controls, and feedback or linear controls. There is also fuzzy logic, which attempts to combine some of the design simplicity of logic with the utility of linear control.

Frequency

Frequency is the number of occurrences of a repeating event per unit time. It is also referred to as temporal frequency. The period is the duration of one cycle in a repeating event, so the period is the reciprocal of the frequency.

Frequency response

Frequency response is the measure of any system's output spectrum in response to an input signal. In the audible range it is usually referred to in connection with electronic amplifiers, microphones and loudspeakers. Radio spectrum frequency response can refer to measurements of coaxial cables, category cables, video switchers and wireless communications devices.

Compensation

In engineering, compensation is planning for side effects or other unintended issues in a design. The design of an invention can itself also be to compensate for some other existing issue or exception.

One example is in a voltage-controlled crystal oscillator (VCXO), which is normally affected not only by voltage, but to a lesser extent by temperature.

Linearization

In mathematics and its applications, linearization refers to finding the linear approximation to a function at a given point. In the study of dynamical systems, linearization is a method for assessing the local stability of an equilibrium point of a system of nonlinear differential equations or discrete dynamical systems. This method is used in fields such as engineering, physics, economics, and ecology.

Estimator

In statistics, an estimator is a rule for calculating an estimate of a given quantity based on observed data: thus the rule and its result (the estimate) are distinguished This is in contrast to an interval estimator, where the result would be a range of plausible values (or vectors or functions).

Block	A block in the context of extraction of petroleum, (and natural gas) signifies a sub-division of an underground reservoir for the purpose of administering a certain production allowable.
Block diagram	Block diagram is a diagram of a system, in which the principal parts or functions are represented by blocks connected by lines, that show the relationships of the blocks. They are heavily used in the engineering world in hardware design, software design, and process flow diagrams. The block diagram is typically used for a higher level, less detailed description aimed more at understanding the overall concepts and less at understanding the details of implementation.
Transfer function	A transfer function is a mathematical representation, in terms of spatial or temporal frequency, of the relation between the input and output of a linear time-invariant system. With optical imaging devices, for example, it is the Fourier transform of the point spread function (hence a function of spatial frequency) i.e. the intensity distribution caused by a point object in the field of view. Explanation Transfer functions are commonly used in the analysis of systems such as single-input single-output filters, typically within the fields of signal processing, communication theory, and control theory.
Case study	A case study is a research methodology common in social science. It is based on an in-depth investigation of a single individual, group, or event. They may be descriptive or explanatory.
Reaction	The third of Newton's laws of motion of classical mechanics states that forces always occur in pairs. Every action is accompanied by a reaction of equal magnitude but opposite direction. This principle is commonly known in the Latin language as actio et reactio.

Nonlinear	In mathematics, a nonlinear system is a system which is not linear, that is, a system which does not satisfy the superposition principle, a nonlinear system is any problem where the variable(s) to be solved for cannot be written as a linear combination of independent components. A nonhomogeneous system, which is linear apart from the presence of a function of the independent variables, is nonlinear according to a strict definition, but such systems are usually studied alongside linear systems, because they can be transformed to a linear system of multiple variables.
Stability	In probability theory, the stability of a random variable is the property that a linear combination of two independent copies of the variable has the same distribution, up to location and scale parameters. The distributions of random variables having this property are said to be "stable distributions". Results available in probability theory show that all possible distributions having this property are members of a four-parameter family of distributions.
Damper	A damper is a valve or plate that stops or regulates the flow of air inside a duct, chimney, VAV box, air handler, or other air handling equipment. A damper may be used to cut off central air conditioning (heating or cooling) to an unused room, or to regulate it for room-by-room temperature and climate control. Its operation can be manual or automatic.
Voltage	The voltage between two points is a short name for the electrical force that would drive an electric current between those points. Specifically, voltage is equal to energy per unit charge. In the case of static electric fields, the voltage between two points is equal to the electrical potential difference between those points.
Complexity	In general usage, complexity tends to be used to characterize something with many parts in intricate arrangement. The study of these complex linkages is the main goal of network theory and network science. In science there are at this time a number of approaches to characterizing complexity, many of which are reflected in this article.
Altimeter	An altimeter is an instrument used to measure the altitude of an object above a fixed level. The measurement of altitude is called altimetry, which is related to the term bathymetry, the measurement of depth underwater. Pressure altimeter Altitude can be determined based on the measurement of atmospheric pressure.

Autopilot	An autopilot is a mechanical, electrical, or hydraulic system used to guide a vehicle without assistance from a human being. An autopilot can refer specifically to aircraft, self-steering gear for boats, or auto guidance of space craft and missiles. The autopilot of an aircraft is sometimes referred to as "George".
Carburetor	A carburetor carburettor, or carburetter (Commonwealth spelling) is a device that blends air and fuel for an internal combustion engine. It is sometimes shortened to carb in North America and the United Kingdom. Word origin The word carburetor comes from the French carbure meaning "carbide".
Ratio	In mathematics, a ratio expresses the magnitude of quantities relative to each other. Specifically, the ratio of two quantities indicates how many times the first quantity is contained in the second and may be expressed algebraically as their quotient. Mathematically, a proportion is defined as the equality of two ratios.
Fuel injection	Fuel injection is a system for mixing fuel with air in an internal combustion engine. It has become the primary fuel delivery system used in automotive petrol engines, having almost completely replaced carburetors in the late 1980s. A fuel injection system is designed and calibrated specifically for the type(s) of fuel it will handle.
Throttle	A throttle is the mechanism by which the flow of a fluid is managed by constriction or obstruction. An engine's power can be increased or decreased by the restriction of inlet gases (i.e., by the use of a throttle), but usually decreased. The term throttle has come to refer, informally and incorrectly, to any mechanism by which the power or speed of an engine is regulated.

Simulink	Simulink, developed by MathWorks, is a commercial tool for modeling, simulating and analyzing multidomain dynamic systems. Its primary interface is a graphical block diagramming tool and a customizable set of block libraries. It offers tight integration with the rest of the MATLAB environment and can either drive MATLAB or be scripted from it.
Random walk	A random walk, is a mathematical formalisation of a trajectory that consists of taking successive random steps. The results of random walk analysis have been applied to computer science, physics, ecology, economics, psychology and a number of other fields as a fundamental model for random processes in time. For example, the path traced by a molecule as it travels in a liquid or a gas, the search path of a foraging animal, the price of a fluctuating stock and the financial status of a gambler can all be modeled as random walks.
Semiconductor	A semiconductor is a material with electrical conductivity due to electron flow (as opposed to ionic conductivity) intermediate in magnitude between that of a conductor and an insulator. This means a conductivity roughly in the range of 10^3 to 10^{-8} siemens per centimeter. Semiconductor materials are the foundation of modern electronics, including radio, computers, telephones, and many other devices.
Manufacturing	Manufacturing is the use of machines, tools and labor to produce goods for use or sale. The term may refer to a range of human activity, from handicraft to high tech, but is most commonly applied to industrial production, in which raw materials are transformed into finished goods on a large scale. Such finished goods may be used for manufacturing other, more complex products, such as aircraft, household appliances or automobiles, or sold to wholesalers, who in turn sell them to retailers, who then sell them to end users - the "consumers".
Pyrometer	A Pyrometer is a non-contacting device that intercepts and measures thermal radiation, a process known as pyrometry. This device can be used to determine the temperature of an object's surface. The word Pyrometer comes from the Greek word for fire, 'πυρ' , and meter, meaning to measure.
Halogen	The halogens or halogen elements are a series of nonmetal elements from Group 17 IUPAC Style (formerly: VII, VIIA) of the periodic table, comprising fluorine (F), chlorine (Cl), bromine (Br), iodine (I), and astatine (At). The artificially created element 117, provisionally referred to by the systematic name ununseptium, may also be a halogen.

	The group of halogens is the only periodic table group which contains elements in all three familiar states of matter at standard temperature and pressure.
Halogen lamp	A halogen lamp, is an incandescent lamp with a tungsten filament contained within an inert gas and a small amount of a halogen such as iodine or bromine. Because of this, a halogen lamp can be operated at a higher temperature than a standard gas-filled lamp of similar power and operating life. The higher operating temperature results in light of a higher color temperature.
Heat transfer	Heat transfer is a discipline of thermal engineering that concerns the transfer of thermal energy from one physical system to another. Heat transfer is classified into various mechanisms, such as heat conduction, convection, thermal radiation, and phase-change transfer. Engineers also consider the transfer of mass of differing chemical species, either cold or hot, to achieve heat transfer.
Linear model	In statistics, the term linear model is used in different ways according to the context. The most common occurrence is in connection with regression models and the term is often taken as synonymous with linear regression model. However the term is also used in time series analysis with a different meaning.
Temperature	Temperature is a physical property of matter that quantitatively expresses the common notions of hot and cold. Objects of low temperature are cold, while various degrees of higher temperatures are referred to as warm or hot. Quantitatively, temperature is measured with thermometers, which may be calibrated to a variety of temperature scales.
Trajectory	A trajectory is the path a moving object follows through space as a function of time. The object might be a projectile or a satellite, for example. It thus includes the meaning of orbit--the path of a planet, an asteroid or a comet as it travels around a central mass.
Cell	A cell in the context of electronic design automation (EDA) is an abstract representation of a component within a schematic diagram or physical layout of an electronic circuit in software. A cell-based design methodology is a technique that enables designers to analyze chip designs at varying levels of abstraction. For example, one designer may focus on the logical function (high-level) and another may concentrate on physical implementation (low-level).

Flagellum	A flagellum is a tail-like projection that protrudes from the cell body of certain prokaryotic and eukaryotic cells, and functions in locomotion. There are some notable differences between prokaryotic and eukaryotic flagella, such as protein composition, structure, and mechanism of propulsion. An example of a flagellated bacterium is the ulcer-causing Helicobacter pylori, which uses multiple flagella to propel itself through the mucus lining to reach the stomach epithelium.
Ligand	In biochemistry and pharmacology, a ligand is a substance that forms a complex with a biomolecule to serve a biological purpose. In a narrower sense, it is a signal triggering molecule, binding to a site on a target protein. The binding occurs by intermolecular forces, such as ionic bonds, hydrogen bonds and van der Waals forces.
Dynamics	In the field of physics, the study of the causes of motion and changes in motion is dynamics. In other words the study of forces and why objects are in motion. Dynamics includes the study of the effect of torques on motion.
Activity	In chemical thermodynamics, activity is a measure of the "effective concentration" of a species in a mixture. By convention, it is treated as a dimensionless quantity, although its actual value depends on customary choices of standard state for the species. The activity of pure substances in condensed phases (solid or liquids) is normally taken as unity.
Laplace transform	In mathematics, the Laplace transform is a widely used integral transform. Denoted $\mathcal{L}\{f(t)\}$, it is a linear operator of a function f(t) with a real argument t ($t \geq 0$) that transforms it to a function F(s) with a complex argument s. This transformation is essentially bijective for the majority of practical uses; the respective pairs of f(t) and F(s) are matched in tables.
Motion	In physics, motion is a change in position of an object with respect to time. Change in action is the result of an unapplied force. Motion is typically described in terms of velocity, acceleration, displacement, and time.

Perspective	Perspective, in context of vision and visual perception, is the way in which objects appear to the eye based on their spatial attributes; or their dimensions and the position of the eye relative to the objects. There are two main meanings of the term: linear perspective and aerial perspective. Linear perspective As objects become more distant they appear smaller because their visual angle decreases.
Internet	The Internet is a global system of interconnected computer networks that use the standard Internet Protocol Suite (TCP/IP) to serve billions of users worldwide. It is a network of networks that consists of millions of private, public, academic, business, and government networks, of local to global scope, that are linked by a broad array of electronic, wireless and optical networking technologies. The Internet carries a vast range of information resources and services, such as the inter-linked hypertext documents of the World Wide Web (WWW) and the infrastructure to support electronic mail.
Magnetic levitation	Magnetic levitation, maglev, or magnetic suspension is a method by which an object is suspended with no support other than magnetic fields. Magnetic pressure is used to counteract the effects of the gravitational and any other accelerations. Earnshaw's theorem proves that using only static ferromagnetism it is impossible to stably levitate against gravity, but servomechanisms, the use of diamagnetic materials, superconduction, or systems involving eddy currents permit this to occur.
Levitation	Levitation is the process by which an object is suspended by a physical force against gravity, in a stable position without solid physical contact. A number of different techniques have been developed to levitate matter, including the aerodynamic, magnetic, acoustic, electromagnetic, electrostatic, gas film, and optical levitation methods. Physics

For levitation on Earth, first, a force is required directed vertically upwards and equal to the gravitational force, second, for any small displacement of the levitating object, a returning force should appear to stabilize it.

Analytics

The simplest definition of Analytics is 'the science of analysis'. A simple and practical definition, however, would be how an entity (i.e., business) arrives at an optimal or realistic decision based on existing data. Business managers may choose to make decisions based on past experiences or rules of thumb, or there might be other qualitative aspects to decision making; but unless there are data involved in the process, it would not be considered Analytics.

Analytic continuation

In complex analysis, a branch of mathematics, analytic continuation is a technique to extend the domain of a given analytic function. Analytic continuation often succeeds in defining further values of a function, for example in a new region where an infinite series representation in terms of which it is initially defined becomes divergent.

The step-wise continuation technique may, however, come up against difficulties.

Region

Region is most commonly found as a term used in terrestrial and astrophysics sciences, notably among the different sub-disciplines of Geography, studied by regional geographers. Regions consist of subregions that contain clusters of like areas that are distinctive by their uniformity of description based on a range of statistical data, for example demographic, and locales. In astrophysics some regions have science-specific terms such as galactic clusters.

Two-sided Laplace transform

In mathematics, the two-sided Laplace transform is an integral transform closely related to the Fourier transform, the Mellin transform, and the ordinary or one-sided Laplace transform. If $f(t)$ is a real or complex valued function of the real variable t defined for all real numbers, then the two-sided Laplace transform is defined by the integral

$$\mathcal{B}\left\{f(t)\right\} = F(s) = \int_{-\infty}^{\infty} e^{-st} f(t)\, dt.$$

The integral is most commonly understood as an improper integral, which converges if and only if each of the integrals

$$\int_0^\infty e^{-st} f(t)\,dt, \quad \int_{-\infty}^0 e^{-st} f(t)\,dt$$

exists. There seems to be no generally accepted notation for the two-sided transform, the $\mathcal{B}$ used here recalls "bilateral".

Signal

In the fields of communications, signal processing, and in electrical engineering more generally, a signal is any time-varying or spatial-varying quantity.

In the physical world, any quantity measurable through time or over space can be taken as a signal. Within a complex society, any set of human information or machine data can also be taken as a signal.

Derivative

In chemistry, a derivative is a compound that is derived from a similar compound by some chemical or physical process. In the past it was also used to mean a compound that can be imagined to arise from another compound, if one atom is replaced with another atom or group of atoms, but modern chemical language now uses the term structural analogue for this meaning - thus eliminating ambiguity of both terms. The term "structural analogue" is common in organic chemistry.

Integral

Integration is an important concept in mathematics and, together with differentiation, is one of the two main operations in calculus. Given a function f of a real variable x and an interval [a, b] of the real line, the definite integral

is defined informally to be the net signed area of the region in the xy-plane bounded by the graph of f, the x-axis, and the vertical lines x = a and x = b.

The principles of integration were formulated independently by Isaac Newton and Gottfried Leibniz in the late 17th century.

Convolution

In mathematics and, in particular, functional analysis, Convolution is a mathematical operation on two functions f and g, producing a third function that is typically viewed as a modified version of one of the original functions. Convolution is similar to cross-correlation. It has applications that include statistics, computer vision, image and signal processing, electrical engineering, and differential equations.

Pole

In the mathematical field of complex analysis, a pole of a meromorphic function is a certain type of singularity that behaves like the singularity of $\frac{1}{z^n}$ at z = 0. This means that, in particular, a pole of the function f(z) is a point a such that f(z) approaches infinity as z approaches a.

Formally, suppose U is an open subset of the complex plane C, a is an element of U and f : U {a} → C is a function which is holomorphic over its domain. If there exists a holomorphic function g : U → C and a positive integer n, such that for all z in U {a}

$$f(z) = \frac{g(z)}{(z-a)^n}$$

holds, then a is called a pole of f.

Solution

In chemistry, a solution is a homogeneous mixture composed of two or more substances. In such a mixture, a solute is dissolved in another substance, known as a solvent. The solvent does the dissolving.

CPSIA information can be obtained at www.ICGtesting.com
Printed in the USA
BVOW050427220812

298417BV00001B/16/P

9 781614 902232